The Transcendent Blueprint And Other Laws Changing Everything We Know

C D Olson

© 2025 CD Olson. All rights reserved. This book or any portion thereof may not be reproduced or used in any manner whatsoever without the express written permission of the publisher except for the use of brief quotations in a book review.

ISBN 979-8-35099-009-6
TXu 2-463-404

Contents

Introduction: A Personal Journey to Universal Truths	vii
Part 1 Continuation of Thought	1
Chapter 1: The Transcendent Blueprint Theory	2
Chapter 2: How Universal Laws Manifest — Context and Variability	12
Chapter 3: The Seed as Proof of Concept	17
Chapter 3.2: The Seed Law	23
Part 2 What Are the Rules	27
Chapter 4: Foundations of Universal Laws	28
Chapter 5: Biology — The Cycles of Life	41
Chapter 6: Physics — The Mechanics of Existence	46
Chapter 7: Thought — The Blueprint of the Mind	58
Chapter 8: Humanity Introduction to the Universal Blueprint	66
Chapter 9: The Gravity of Words	72
Chapter 10: Humanity — An Advanced Look at the Living Blueprint	89
Part ∞ The X Chapters - Rethinking Thought	107
Chapter X: Neutrality and Freedom	108
Chapter XX: Dark Matter and the Blueprint of Balance	112
Chapter XXX: The Seed of the Universe	119
Chapter XL: Rethinking Time Through the Blueprint of Growth	126
Chapter L: Rethinking Evolution Through the Blueprint of Time	134
Chapter LX: The Law of Work	151
Chapter LXX: The Law of Friction	156
Chapter LXXX: The Flow of Alignment: Spirit, Energy, and the Blueprint	161
Chapter LXXXI: The Law of Aligned Causality	166

Part 3 Interpretations 174

    Chapter 11: Faith — Transcendence Beyond Belief 175

    Chapter 12: The Carbon Question 190

    Chapter 13: Cancer Through the Lens of the

    Transcendent Blueprint Theory (TBT) 203

    Chapter 14: The Womb as a Blueprint for Life 209

Part 4 The Future 213

    Chapter 15: The Unique Promise of TBT 214

    Chapter 16: The Final Reflection 219

Part 5 The End 223

Olson's Theories, Laws, and Hypotheses Featured in This Book

1. **Transcendent Blueprint Theory** *(Intro, Ch 1, 2)*
2. **The Seed Law** *(Chapter 3.2)*
3. **Dark Matter Hypothesis** *(Chapter XX)*
4. **The Seed Cosmos Hypothesis** *(Chapter XXX)*
5. **The Contextual Time Hypothesis** *(Chapter XL)*
6. **The Transcendent Blueprint Theory of Change** *(Chapter L)*
7. **The Law of Work** *(Chapter LX)*
8. **The Law of Friction** *(Chapter LXX)*
9. **The Law of Aligned Causality** *(Chapter LXXXI)*

1. Transcendent Blueprint Theory (Intro, Ch 1, 2)

This is entirely my own, as it synthesizes universal principles (cycles, reciprocity, transformation, etc.) into a cohesive framework that bridges disciplines like biology, physics, and sociology. While others may have addressed universal principles, my specific articulation and integration is unique.

2. The Seed Law (Chapter 3.2)

The Seed Law, as I've defined it—encoded potential within origins driving growth and transformation—is an original framing. It draws inspiration from natural and universal processes but positions them in a way that integrates biological, conceptual, and physical systems under TBT.

3. Dark Matter Hypothesis (Chapter XX)

This explores dark matter through TBT principles, presenting a unique angle. While dark matter is a widely discussed topic in physics, my interpretation through the lens of balance, transformation, and connectivity is novel.

4. The Seed Cosmos Hypothesis (Chapter XXX)

The idea of the cosmos itself as a seed, encoding potential for universal expansion and transformation, is an original metaphor and theoretical perspective. It connects cosmology to universal laws in a way that reflects the broader framework.

5. The Contextual Time Hypothesis (Chapter XL)

This hypothesis is distinct, particularly reframing time as a growth-based, contextual phenomenon. While others have explored nonlinear or relative time, my application through the TBT lens offers a fresh interpretation.

6. The Transcendent Blueprint Theory of Change (Chapter L)

My theory linking change to cycles, reciprocity, and transformation under TBT principles and Contextual Time is a new conceptualization, providing an alternative to traditional evolutionary theories.

7. The Law of Work (Chapter LX)

The Law of Work asserts that no creation, transformation, or outcome occurs without the expenditure of energy, effort, or intentional action. This principle, while reflecting universal truths observed in physics, biology, and human systems, is uniquely positioned within TBT as the driving mechanism for all progress. Work transforms potential into reality, aligning systems with purpose and growth.

8. The Law of Friction (Chapter LXX)

The Law of Friction reveals that resistance, when harnessed, becomes a tool of refinement and transformation. It demonstrates that obstacles, challenges, and opposition are necessary forces within all systems, compelling adaptation, growth, and balance. My framing of friction as a dual catalyst—both a constraint and a driver—is a unique contribution that integrates TBT principles into physical, biological, and conceptual systems.

9. The Law of Aligned Causality (Chapter LXXXI)

This law, focusing on causality within aligned systems to ensure sustainable outcomes, is an original contribution. While causality is a broad philosophical and scientific concept, my framing under TBT principles makes it unique.

Introduction: A Personal Journey to Universal Truths

One of my earliest memories is sitting in a snowbank, gazing intently at an object in the distance. My focus was so absolute that the world around me seemed to blur into something surreal—a kaleidoscope of particles dancing in and out of existence. At the time, I dismissed this experience as an overactive imagination or perhaps a lack of eating enough carrots. But now, decades later, I see it for what it truly was: the seed of a realization that has grown with me over the years.

To understand this memory and what it represents, it's important to know a little about me. I was raised for 3 years by my dual diagnosed paranoid schizophrenic mother, which was a time of immense unpredictability that shaped my earliest perceptions of the world. From a young age, I knew I was different. I saw things differently, sought out the truth and found comfort in deep thought. That snowbank memory was post that time, having been removed from the environment, and references the beginning of my journey of deep introspection—a place where I believed, as a child, that if I could just think hard enough, I could move objects or alter reality itself. I never moved an object, but I stumbled upon something even greater: the foundation of a conceptual framework that would take me 50 years to fully realize.

This book introduces the *Transcendent Blueprint Theory*, an idea born of decades of observation, reflection, and study. Its roots stretch back to my childhood but came into sharper focus during a time of personal struggle—when I was near delirious, sick with pneumonia, seeking clarity from God. I asked a question, and in that moment of sickness, was given an answer. For a moment, I was allowed to glimpse something extraordinary: a blueprint underlying the universe, connecting everything from biology to physics, from human thought to the movement of carbon. It was as if a door had opened in my mind, revealing patterns and principles that I had been piecing together my entire life. Every question I had now had an answer, and the answers all started with the *Transcendent Blueprint Theory*.

The *Transcendent Blueprint Theory*: A Unifying Vision

At its core, the *Transcendent Blueprint Theory* **(TBT)** reveals that universal principles—**cycles, reciprocity, motion, connection, and transformation**—govern every aspect of existence. These laws are not confined to one discipline; they bridge **physics, biology, psychology, sociology**, and even the **abstract systems** of thought, culture, and faith. They guide the movements of galaxies as much as they shape the growth of a tree, the flow of ideas, and the choices we make in our daily lives.

This book will challenge you to **see the world differently**—to recognize the patterns underlying everything from carbon cycles to human relationships. TBT provides a lens to understand **why systems thrive, collapse, and renew**—whether they are ecosystems, civilizations, or even your personal life.

What You Will Discover

Through this journey, we will explore groundbreaking insights that transform how we see **science, society, faith**, and the very fabric of existence. The *Transcendent Blueprint Theory* reveals how universal principles connect systems as vast as the cosmos and as intimate as human thought. Here's what you will uncover:

1. **Carbon Seen Through the Blueprint of Balance**
   - Carbon is not a villain but a **mediator of balance, motion, and transformation**. By reframing environmental challenges, we recognize carbon's role in growth, renewal, and cycles, transforming it into part of the solution rather than the problem.

2. **Words Through the Lens of Physics**
   - Words are not merely abstract sounds—they carry **mass, motion, and energy**. Like physical forces, words create friction, momentum, and magnetism, shaping the realities we inhabit.

3. **Faith as a Reflection of Universal Laws**
   - Spiritual teachings about **cycles, reciprocity, and transformation** align with the universal principles revealed in TBT. Faith and science are not opposites but partners, reflecting the same transcendent truths that connect all systems.

4. **People and Relationships Through Magnetism**
   - Human relationships operate on unseen forces, mirroring the **attraction and repulsion** seen in magnetic fields. Understanding this alignment sheds light on **connection, conflict, and growth** in individuals and societies.

5. **Thought Mirrored in Biological Processes**
   - Creativity and thought mirror biological processes like **cellular growth and renewal**. The mind, like nature, follows cycles of **adaptation, decay, and transformation**.

6. **The Cosmos as a Seed: The Blueprint of Expansion**
   - The universe itself operates like a seed, carrying **encoded potential for growth and transformation**. Galaxies expand, stars change, and even dark matter reflects the blueprint of balance and connection, showing that the same principles govern all of creation.

7. **Time Through Context and Growth**
   - Time is not linear; it is **contextual**, shaped by growth, cycles, and perspective. The **Contextual Time Hypothesis** reveals how systems experience time differently, from the rapid expansion of the cosmos to the gradual unfolding of human lives.

8. **Adaptation as an Expression of Universal Cycles**
   - Evolution is not random change but rather **transformation and adaptation within created kinds**, guided by the same cycles of **motion, resistance, and transformation** that shape all systems. Organisms, like societies and ideas, respond to challenges by aligning with natural laws encoded in their DNA, ensuring balance and continuity.

9. **Humanity as a Living System**
   - Humanity behaves like a living organism, competing for space, resources, and ideological dominance. Misalignment leads to stagnation or decline, but understanding humanity through the blueprint reveals pathways for **renewal, balance, and growth**.

10. **Motion, Friction, and Transformation in Societal Progress**
    - Resistance and conflict are not failures but essential forces of **refinement**. Like friction that smooths a stone or tension that generates energy, societal challenges drive progress and adaptation, ensuring systems evolve rather than stagnate.

11. **The Universal Connection Between All Systems**
    - TBT shows that no system exists in isolation. The same principles of **reciprocity, connection, and balance** govern

ecosystems, human relationships, and even galaxies. These underlying laws hold the key to **understanding and aligning** our actions with the blueprint.

A Blueprint for Everything

The *Transcendent Blueprint Theory* will challenge you to see the world differently:

- **The Cosmos** reveals the encoded patterns of expansion, balance, and potential.
- **Time** is reframed as a contextual force, shaped by cycles of growth and perspective.
- **Evolution** not evolution at all but transformation or adaptation within created kinds originating from seeds with imprinted DNA.
- **Humanity** is shown as a living organism that thrives when aligned with universal laws.

From the flow of carbon to the energy of thought, from the smallest biological processes to the vast movements of galaxies, TBT uncovers the hidden connections that unite all existence.

Humanity Through the Blueprint Lens

One of TBT's most profound revelations lies in its reframing of **humanity itself**. On the surface, human behavior appears random and chaotic. But through this blueprint, patterns emerge that align with **universal laws**:

- **Growth and Collapse**: Human civilizations follow **cyclical patterns**, much like ecosystems. Unchecked growth leads to chaos; resistance sparks refinement; and collapse often fuels renewal.
- **Free Thought and Motion**: When thought is stifled, systems stagnate and decline, much like life when deprived of oxygen. Resistance and friction, however, drive growth and adaptation, ensuring forward motion.
- **Connection and Reciprocity**: Societies, like ecosystems, thrive on the **exchange of resources, ideas, and energy**. Attempts to manipulate or suppress this balance destabilize systems and create voids.

- **Magnetism in Relationships**: Human interactions mirror magnetic forces—**attraction, repulsion, and alignment**—revealing unseen laws that govern connection and conflict.

By viewing humanity as part of a larger, interconnected system, TBT provides clarity amidst the chaos. It shows that our struggles—personal, societal, and global—stem not from randomness but from **misalignment with universal principles**.

A Blueprint for Transformation

The *Transcendent Blueprint Theory* does not merely explain humanity or nature—it transforms our approach to solving **real-world challenges**:

- **Reframing Environmental Crises**: Seeing carbon and climate through the correct lens reveals solutions that work in harmony with natural cycles, not against them.
- **Unlocking Thought and Creativity**: Understanding thought as energy allows us to **harness its motion**, unlocking new levels of innovation and progress.
- **Bridging Science and Faith**: TBT shows how spiritual truths align with scientific principles, offering a unifying perspective for understanding the world.
- **Restoring Balance to Human Systems**: By aligning with principles of motion, reciprocity, and connection, societies can rediscover the pathways to **growth, balance, and renewal**.

A Call to Curiosity

What if the same forces that govern stars and galaxies also guide our relationships, thoughts, and cultures? What if solving the mysteries of carbon and motion could unlock not just environmental solutions but a deeper understanding of **humanity's purpose and potential**?

This book is an invitation to see the world through the lens of **universal connection**—to recognize that every system, whether seen or unseen, operates under the same laws. It is not merely an exploration of physics, biology, or thought but a **revelation of the blueprint** that connects them all.

The Transcendent Blueprint Theory will help you:

- **Understand the patterns** that govern the universe and humanity.

- **Reframe challenges** as part of natural cycles of growth, decay, and renewal.
- **Unlock clarity and transformation** in your personal life, your work, and the world around you.

Welcome to the Blueprint

This is more than a theory—it is a **roadmap** for understanding our place in the universe. It bridges the seen and unseen, the scientific and spiritual, offering a way forward for individuals, societies, and humanity as a whole.

The same laws that sustain the cosmos govern our lives. If we understand, apply and align ourselves with these principles, we stand on the edge of a future of change.

Welcome to the Transcendent Blueprint Theory. Let your journey begin.

Part 1
Continuation of Thought

Chapter 1: The Transcendent Blueprint Theory

Beneath the surface of every system—natural, human, or cosmic—there exists an interconnected framework that governs all existence. This framework is not confined by discipline or scale but operates universally. From the cycles of nature to the ripple of ideas, from the magnetic pull of relationships to the flow of energy within and between us, this framework reveals the unity of all things.

I call this framework the Transcendent Blueprint Theory. It is a universal model that uncovers the patterns and laws that govern life, thought, and matter. These principles—cycles, reciprocity, transformation, motion, and connection—transcend disciplines, revealing the connections between seemingly unrelated phenomena. The theory shows us how the same principles that guide the growth of a tree also shape human thought, relationships, and global systems.

This chapter lays the foundation for understanding the *Transcendent Blueprint Theory*, presenting the universal laws that shape the world around us and within us. It invites you to see the connections that bind life, thought, and energy into a unified whole.

A Concise Definition of the Transcendent Blueprint Theory

The Transcendent Blueprint Theory (TBT) is a groundbreaking framework that reveals the immutable principles—**cycles, reciprocity, transformation, motion, and connection**—that govern all existence. Unlike existing theories that focus on isolated disciplines, TBT uniquely integrates these principles across natural, human, and abstract systems, demonstrating that all things are inherently subject to these universal laws and cannot operate outside them. It transcends boundaries, offering a unified lens for understanding the interconnectedness of thought, action, and matter. Rooted in the understanding that everything functions within these immutable truths, TBT challenges humanity's perception of control by presenting a model for observing, understanding, and harmonizing with these principles. Its originality lies in its integrative approach: **connecting science, philosophy, and faith** to solve complex problems, identify misalignments, and guide humanity toward sustainable progress, all while honoring the unchanging truths that sustain and connect all systems.

The Universal Blueprint: A Map of Connection

Imagine the universe as a vast web, where each strand represents a different domain of life—physics, biology, thought, or society. At first glance, these strands seem separate, each with its own rules. But closer inspection reveals the same underlying patterns that guide them all.

- **Cycles**: From the turning of seasons to the rise and fall of civilizations, cycles govern growth, decay, and renewal. They ensure continuity, balancing moments of expansion and contraction across all systems.
- **Reciprocity**: The exchange of energy, resources, and support sustains ecosystems and relationships alike. Whether it's the give-and-take of nature or the dynamics of human interaction, reciprocity ensures harmony.
- **Transformation**: Energy and matter are never lost; they shift from one form to another, fueling creation, renewal, and adaptation. This law governs everything from the life cycle of carbon to the evolution of thought.
- **Motion and Magnetism**: Attraction, repulsion, and motion define relationships at every level, from atomic bonds to human connections. Movement creates change, and magnetism draws systems together or drives them apart.
- **Connection**: At its core, the web of our existence is held together by connection. This principle shows up and binds all systems into cohesive wholes creating interdependence between components. From neural networks to ecosystems, connection ensures that nothing functions in isolation.

These principles form the universal blueprint, transcending their contexts to reveal the shared foundations of all systems. By understanding them, we gain a lens through which to see the unity of existence and the forces that govern it.

Newton's Laws of Motion

Newton's laws describe how forces act on objects, revealing patterns of **motion, balance, and resistance**. Through TBT, these patterns become universal principles that govern **all systems**—whether physical, biological, or even social:

- **First Law (Inertia):** *An object in motion stays in motion unless acted upon by an external force.*
- **TBT Connection:** Inertia represents stability and momentum, showing that systems tend to maintain their state until influenced by an outside force. This principle can be seen in ecosystems, economies, and even ideas: a healthy forest, a growing market, or a societal trend will persist unless disrupted. TBT expands this law beyond objects to all forms of movement—physical, energetic, and conceptual—demonstrating how stability and disruption shape outcomes.
- **Third Law (Action and Reaction):** *For every action, there is an equal and opposite reaction.*
- **TBT Connection:** This law highlights the universal balance between cause and effect. In TBT, this principle explains the interconnected nature of all systems: every action triggers a response that ripples through the environment. For example, removing a tree affects soil health, water retention, and nearby organisms. Similarly, a positive or negative action within a social system creates measurable consequences, maintaining balance through an unseen, but fundamental, order.

Einstein's Theory of Relativity

Einstein's work redefined our understanding of time, motion, and space, illustrating their interdependence. Through TBT, relativity aligns with the principle of **connection**:

- **TBT Connection:** Einstein showed that space, time, and energy do not exist independently but are woven together into a unified system. TBT extends this idea, revealing that no system—whether physical, biological, or conceptual—exists in isolation. The interconnected fabric of reality ensures that motion, energy, and even thought ripple across systems, influencing outcomes on multiple scales.

The Power and Position of TBT

The **Transcendent Blueprint Theory (TBT)** does not simply unify tangible and intangible systems under shared principles. Its true strength lies in its ability to **redefine and reframe** established scientific concepts by uncovering their deeper alignment within universal patterns. TBT shows that the same foundational principles—**cycles, transformation, connection, motion, and balance**—govern not just isolated systems but **everything**: from measurable forces like time and magnetism to broader, seemingly abstract phenomena like thought and evolution.

TBT challenges the boundaries of established science and expands its reach:

1. **Rewriting Time:**
    - Traditional science views time as a linear or relative construct (e.g., Einstein's relativity).
    - **TBT Expansion:** Time is not merely a measurement—it's a **contextual system** driven by cycles, transformation, and motion. Time flows differently across scales and systems, influenced by energy, growth, and decay.

2. **Reframing Evolution:**
    - Evolution is often described as a process driven by random mutation and natural selection.
    - **TBT Expansion:** Evolution is not random; it is guided by universal principles, particularly those outlined in the **Seed Law**. Every system contains encoded potential at its origin (the "seed") that governs its trajectory. Systems adapt for **balance**, **efficiency**, and **connection**, following predictable patterns of transformation and alignment.
    - When combined with the **Seed Law** and the **Contextual Time Theory**, mainstream interpretations of evolution—based on randomness and chance—no longer hold up. Contextual time reveals that systems evolve in response to their environment on timelines influenced by energy flow, scale, and purpose. Evolution is, therefore, an expression of intentional, universal patterns rather than a product of blind chance.

3. **Reexploring Carbon and Magnetism:**
    - Carbon and magnetism are measurable forces in physical science, fundamental to life and motion.
    - **TBT Expansion:** These forces are not isolated—they represent universal principles in action. Carbon cycles embody transformation and reciprocity. Magnetism reflects motion,

friction, and the interplay of opposites, principles that extend far beyond the physical into thought, relationships, and even societal forces.

Tangible and Intangible Systems Revisited

While TBT's strength lies in its ability to **reframe existing laws**, it also bridges the gap between systems:

- **In tangible systems**—like forests, water cycles, and physics—TBT deepens our understanding of how universal laws operate at every scale.
- **In intangible systems**—like thought, societal structures, and culture—TBT provides a way to identify and apply these same universal principles.

Why This Matters

The power of the *Transcendent Blueprint Theory* is not just in its ability to unify, but in its capacity to **transform our understanding of everything**:

- Scientific laws are no longer isolated to physical phenomena.
- Abstract concepts are no longer vague or immeasurable.

By recognizing the universal principles at play—whether in the motion of planets, the flow of carbon, the adaptation of species, or the magnetism of human behavior—TBT gives us a new lens to see reality. It is a framework that reveals **order, purpose, and the shared patterns** that govern all systems, measurable and immeasurable, physical and conceptual.

For Example:

Newton's Laws of Motion

Newton's laws traditionally describe the behavior of physical objects, but under TBT, they reveal deeper insights into human behavior and systems:

- **First Law (Inertia):** *An object at rest stays at rest unless acted upon by an external force.*
- **TBT Application:** Stagnation in life or work mirrors physical inertia—progress requires an external force. Initiating action, like pursuing meaningful projects or expanding your network, generates momentum and breaks the cycle of stillness.

- **Third Law (Action and Reaction):** *For every action, there is an equal and opposite reaction.*
- **TBT Application:** Efforts, whether time, energy, or resources, produce returns, much like planting seeds in nature. This reflects **reciprocity**, where every input generates a proportional response, balancing systems across all levels.

Einstein's Theory of Relativity

Relativity explains the relationship between time, space, and motion, but TBT expands its scope to new contexts:

- **Context and Scale:** Relativity shows that systems are influenced by their surroundings and scale. For example, a small act of kindness can ripple across relationships, magnifying its impact over time—similar to how subtle shifts in space-time influence cosmic events.
- **Abstract Systems:** Relativity applies to emotional and societal dynamics, where outcomes (like connection or growth) depend on the perspective and position of participants within a dynamic system.

The Second Law of Thermodynamics

The Second Law explains the natural tendency of energy to move toward entropy (disorder). TBT connects this principle to broader, observable systems:

- **Energy Redistribution in Ecosystems:** In a forest, energy dissipated by decaying matter (entropy) fuels the growth of new plants and organisms. This reflects cycles of **renewal** and **transformation** inherent in TBT.
- **Cross-Application in Physics and Biology:** Energy dissipation (e.g., heat loss in a closed system) mirrors nutrient recycling in soil. Heat released during decay fuels microbial activity, enriching the ecosystem and perpetuating the cycle of life.
- **Bridging Laws:** Entropy also connects to Newton's Third Law (Action and Reaction). As energy disperses, its reaction often triggers renewal in another system. For instance, sunlight

absorbed by plants (thermodynamics) sets photosynthesis in motion, creating the foundation for life.

Expanding the Concept

The transcendence of these laws can also be observed in society, psychology, and human behavior. For instance:

- **Societal Systems:** Periods of disorder (*entropy*) often precede transformation and renewal. Historical revolutions or social movements demonstrate how breakdowns in systems can catalyze growth and progress.
- **Psychology:** Emotional entropy—stress, chaos, or mental fatigue—can act as a precursor to breakthroughs. When properly processed and balanced, it often leads to personal growth and clarity.
- **Human Behavior:** The energy "dissipated" in one pursuit often fuels growth in another. For example, effort spent overcoming failure or mastering a challenge contributes to cycles of learning, renewal, and resilience.

Why This Matters

By reframing scientific laws through the lens of TBT, we reveal their broader relevance across disciplines. The same principles that govern **physical systems**—like energy, motion, and cycles—also shape **intangible systems**, such as relationships, thought, and culture.

This expanded understanding creates a bridge between the tangible and the abstract, offering deeper insights into the patterns that guide both **science** and **humanity**. TBT does not just unify—it empowers us to apply universal principles practically and meaningfully in all areas of life.

Aligning Discoveries Under Universal Principles

Humanity has always sought to understand the rules that govern existence. Each discovery—Newton's, Einstein's, and beyond—has expanded our knowledge of specific phenomena. The TBT offers a different lens, not to replace these discoveries but to show how they align under universal principles. By recognizing the connections between truths, principles, and laws, we move closer to understanding the blueprint that sustains all forms of existence.

Recap: The Power of TBT

The *Transcendent Blueprint Theory* **(TBT)** unifies what traditional thinking keeps separate, revealing the shared principles that connect all systems—tangible and intangible. By recognizing these patterns, we break free from compartmentalized thinking and discover:

- **Shared Patterns**: The growth of a tree reflects the same principles we observe in larger systems, such as the cycles and expansion of the cosmos.
- **Cross-Disciplinary Influence**: The laws of physics, biology, and thought interact and inform one another. Motion shapes human progress, biology influences societal patterns, and cycles govern thought and behavior.
- **Alignment**: When we recognize and live in harmony with universal principles like cycles, motion, and transformation, we create balance, growth, and sustainability in both personal and collective systems.

By understanding this blueprint, we move beyond limitations, gaining a framework to observe, connect, and act intentionally across all areas of life.

To fully unlock TBT's potential, we must examine its **scientific foundation** and **core tenets**—the universal principles that govern all systems and provide the structure for this transformative framework.

Core Tenets of the Transcendent Blueprint Theory

The theory is built on five foundational principles that reveal universal laws operating across all systems—physical, biological, and conceptual:

1. **Cycles: All systems operate within cycles of growth, decay, and renewal. Cycles ensure continuity, transformation, and stability across time and scale.**
2. **Transformation: Change is constant and inevitable. Systems evolve through adaptation, energy flow, and intentional or natural shifts that drive progress.**
3. **Motion and Friction: Movement initiates transformation, while friction creates resistance that refines systems, acting as a catalyst for growth and balance.**

4. **Connection**: No system exists in isolation. Systems are interconnected through relationships, energy flow, and mutual influence, creating larger patterns of unity.

5. **Reciprocity and Balance**: Systems maintain equilibrium through reciprocal exchanges—inputs create outputs, actions generate responses. This balance sustains stability and harmony across all domains.

Why It Matters

These tenets form the foundation of the **Transcendent Blueprint Theory**, illustrating that universal principles are not confined to isolated disciplines. By understanding and applying these laws, we align with the natural rhythms of existence, fostering intentional progress and sustainable growth in all areas of life.

How the Theory Applies

The **Transcendent Blueprint Theory (TBT)** serves as a lens to uncover the universal principles shaping seemingly disparate systems, from natural phenomena to abstract concepts. By connecting **cycles, reciprocity, transformation, motion, and magnetism** across disciplines, TBT reveals:

- **The Laws of Nature**: How universal principles sustain ecosystems, drive balance, and create resilience within biological and physical systems.
- **The Framework of Thought**: How ideas, beliefs, and emotions follow the same cycles of growth, decay, and renewal seen in nature, evolving through reciprocal exchange and transformation.
- **The Mechanics of Relationships**: How human connections reflect patterns of motion and magnetism, illustrating forces of attraction, repulsion, and collaboration that shape interactions.
- **Humanity as a System**: How humanity operates within the blueprint, where actions ripple across environmental, social, and cultural systems. Aligning with universal principles offers a pathway to restore balance and harmony.
- **The Foundation for New Understanding**: How TBT made it possible to **rethink and expand** core scientific concepts—like time, evolution, carbon, and magnetism—by revealing their

alignment within broader universal patterns. This blueprint challenges limitations of conventional science and reframes these ideas with deeper clarity and purpose.

- **Universal Insight**: How these principles extend beyond the physical to conceptual realms, illuminating topics such as the interplay between science and faith, the energetic nature of words, and humanity's place within the cosmos.

Each chapter explores these applications in detail, demonstrating how the *Transcendent Blueprint Theory* not only connects disciplines but also provides actionable insights to solve problems, align systems, and foster progress that respects the natural rhythms of existence.

The Invitation to See Differently

The *Transcendent Blueprint Theory* is more than a framework—it is a **transformative lens** for understanding the world and humanity's role within it. TBT reveals the universal patterns that connect physical, mental, and emotional realms, guiding us to act with purpose by aligning with principles that sustain growth, balance, and harmony.

This journey is not just about understanding the blueprint—it is about unlocking its potential to **revolutionize** the way we approach the challenges of our time. From advancing scientific discovery to reshaping societal systems, TBT offers a **pathway** to uncover new truths, foster innovation, and inspire solutions that harmonize with the rhythms of life itself.

By the end of this book, you will see how these universal principles shape **every aspect of existence**—from personal choices to global systems. You will discover how aligning with them can **transform perspectives**, inspire breakthroughs, and illuminate paths toward progress.

Let us embark on this journey together—exploring the blueprint that connects us all and opening the door to a future of profound insight and meaningful change.

Chapter 2: How Universal Laws Manifest — Context and Variability

1. Introduction: The Appearance of Inconsistency

If the principles of the Transcendent Blueprint Theory are universal, why do they not appear in all systems in the same way? Why does a tree grow and reproduce while a rock or a piece of metal appears static? Why do thoughts evolve and ripple outward, while inanimate objects seem unchanging?

The answer lies not in a lack of universality but in the contextual expression of these laws. Universal principles like cycles, reciprocity, motion, and transformation govern all systems, but their manifestation depends on the nature, purpose, and context of each system. This variability is not a flaw but a testament to the adaptability of universal laws to operate across disciplines and scales.

Moreover, what we perceive as uneven manifestation is often a matter of perspective. The erosion of a rock operates on a timescale far beyond human perception, while transformations within an atom occur on a scale too small for direct observation. The universal principles are always present; their visibility depends on the tools and perspectives we bring to them.

2. Universal Laws Apply to All Systems

The Transcendent Blueprint Theory posits that all things—living, inanimate, or abstract—are governed by the same universal principles:

- **Cycles**: Patterns of repetition and renewal, present in ecosystems, planetary motion, chemical processes, and even thought patterns.
- **Reciprocity**: The exchange of energy, resources, or influence, evident in biological symbiosis, chemical reactions, and human relationships.
- **Motion and Resistance**: The forces driving change and refinement across all systems.
- **Transformation**: The continual change of form, function, and state, from the cellular level to societal structures.
- **Connection**: The links binding all systems together, ensuring no element exists in isolation.

While these laws are universal, their expression is shaped by the unique characteristics of each system.

3. Contextual Manifestation of Universal Laws

A. The Nature of the System

Every system has its own composition, structure, and role, which influence how universal laws are expressed:

- **Living Systems**: Designed for growth, reproduction, and adaptation, these systems visibly demonstrate cycles, reciprocity, and transformation.
  - *Example:* A tree grows, sheds leaves, and regenerates through energy cycles.
- **Static Systems**: Inanimate systems, such as rocks or metals, express universal laws in subtler, less dynamic ways.
  - *Example:* A rock erodes over time, revealing the influence of cycles and transformation.

Key Insight: Universal laws are always present, but their visibility depends on the system's purpose and design.

B. Energy Flow and Interaction

The degree to which energy flows through a system determines how actively universal laws are expressed:

- **Dynamic Systems**: Living organisms, with constant energy exchange through metabolism, exhibit visible processes like growth, division, and renewal.
- **Static Systems**: Inanimate objects interact with energy passively or reactively, leading to slower or less visible changes.

Key Insight: Systems with high energy flow demonstrate universal laws more dynamically, but even passive systems adhere to these principles.

C. Scale and Complexity

The scale and complexity of a system influence how universal laws manifest:

- **Simple Systems**: Systems like a single atom or a grain of sand follow cycles and transformations but lack the complexity to show interconnected reciprocity or adaptability.

- **Complex Systems**: Ecosystems, human societies, and neural networks demonstrate layered expressions of universal principles, amplifying their visibility.

Key Insight: Complexity amplifies the manifestation of universal laws, while simplicity presents these laws in more fundamental forms.

D. Adaptation as a Dynamic Factor
Adaptation enhances the visibility of universal laws in systems:

- **Living Systems**: Adaptation is a hallmark of life, enabling systems to change and align with environmental demands.
  - *Example:* Cells divide to repair damage, and organisms adapt to changing conditions.

- **Inanimate Systems**: While less adaptive, they still transform under external forces, adhering to principles like motion and transformation.
  - *Example:* A glacier shifts and reshapes landscapes over time.

Key Insight: Adaptability brings dynamic expression to universal laws, making them more apparent in living systems.

4. How Scientific Laws Reflect Contextual Manifestation

The **Transcendent Blueprint Theory (TBT)** aligns with and extends established scientific laws by showing how **universal principles manifest differently based on context**—factors like scale, energy flow, and purpose influence their expression.

Newton's Laws of Motion

Newton's laws describe the relationship between forces, motion, and resistance, but their manifestation depends on context:

- **First Law (Inertia):** *An object in motion remains in motion unless acted upon by an external force.*
- **TBT Context:** In large-scale systems like planets, motion persists almost indefinitely due to minimal resistance. However, in smaller systems—like ecosystems or social dynamics—external forces (environmental changes, decisions) more readily disrupt stability and refine outcomes.

- **Third Law (Action and Reaction):** *For every action, there is an equal and opposite reaction.*
- **TBT Context:** The scale and energy of the action determine the manifestation of the reaction. For example, physical motion produces immediate and measurable responses, while actions in thought or relationships ripple outward with less immediate, but still measurable, impacts.

Einstein's Theory of Relativity

Einstein demonstrated that motion, time, and space are interdependent, but these principles do not manifest uniformly:

- **TBT Context:** Time flows differently depending on energy and scale. For example, relativity applies visibly to celestial systems, but at smaller scales (e.g., human experiences), time is shaped by contextual perception—such as faster pacing in adulthood compared to childhood.

The contextual manifestation of universal principles explains why laws like motion or time do not appear equally in all systems. Scale, energy, and purpose influence how these principles express themselves—what is observable in one system may exist in a subtler or less measurable form in another.

5. Why Contextual Manifestation Strengthens the TBT

A. Universality in Context

The variability in manifestation shows that universal laws transcend rigid boundaries, adapting to the context of each system. This flexibility underscores their universality rather than undermining it.

B. Recognizing Patterns Amid Diversity

By observing how these principles manifest differently in biology, physics, thought, and society, we deepen our understanding of their universality and uncover connections between seemingly unrelated systems.

Key Insight: The adaptability of universal laws is not a limitation but a strength, revealing their profound relevance across all forms of existence.

6. Conclusion: Universality Beyond Boundaries

The *Transcendent Blueprint Theory* demonstrates that universal laws govern all systems, but their manifestations are shaped by the system's nature, energy flow, and complexity.

From the growth of a tree to the erosion of a mountain, from the division of a cell to the evolution of a thought, these principles remain constant. The variation in how they appear reflects not inconsistency but adaptability—proof of their universality.

By understanding these patterns, we gain a lens to see the connections between all things, reminding us that even the most static systems are part of a dynamic, interconnected whole.

Chapter 3: The Seed as Proof of Concept

Introduction: The Seed as a Model of Universality

Every seed carries the extraordinary potential for transformation. Encased within its shell lies a complete blueprint for its future—a tree, a forest, or even an ecosystem. This small, seemingly simple vessel encapsulates the principles that govern all systems, serving as both a tangible example and a theoretical proof for the Transcendent Blueprint Theory (TBT).

This chapter explores the seed not just as a metaphor but as a practical model of the TBT's principles. By dissecting its journey—from dormancy to germination, growth, and systemic influence—we reveal how universal laws like cycles, reciprocity, transformation, and connection operate across all domains of existence.

Section 1: What Is a Seed?

A seed is far more than a dormant object. It is a vessel of stored potential, encoded with the blueprint for what it will become. Its journey reflects the universal principles of transformation and connection to its environment.

Encoded Blueprint

The DNA within a seed preordains its form and function, aligning it with the TBT principle that systems carry within them the instructions for their development.

- *Example:* An acorn contains the encoded potential for an oak tree, with roots, branches, and leaves pre-programmed within its structure.

Stored Potential Energy

The energy within a seed mirrors the conservation of energy in physics, awaiting transformation from potential to kinetic as the seed begins to grow.

- *Example:* Nutrients stored within a seed fuel its germination, demonstrating how dormant energy is transformed into dynamic growth.

Activation Through Interaction
A seed's transformation depends on its relationship with external forces like sunlight, water, and soil. This reliance on external catalysts underscores the TBT principle of connection and reciprocity.

> • *Example:* Photosynthesis transforms sunlight into glucose, fueling the seed's growth, much like systems thrive on reciprocal exchanges of energy and resources.

A seed's ability to carry its entire future within itself while requiring external conditions to activate its growth encapsulates the principles of cycles, reciprocity, and transformation. It bridges the gap between potential and reality, illustrating how universal laws guide systems toward fulfillment.

Section 2: Interdisciplinary Observations of the Seed
The seed is not only a biological phenomenon; it reflects universal principles that transcend disciplines. By exploring the seed's journey, we uncover its relevance to biology, physics, sociology, psychology, and economics.

Biology
The seed serves as the foundation of ecosystems, embodying cycles of life and renewal. It interacts with its environment, giving and receiving energy and resources to sustain life.

> • *Example:* A seed grows into a tree that stabilizes soil, provides oxygen, and offers habitat, creating a reciprocal exchange that sustains the ecosystem.

Physics
The seed's transformation mirrors the energy flows and cycles observed in the physical universe. The potential energy stored within a seed reflects the encoded blueprint that drives systems toward transformation and expansion.

> • *Example:* Just as the universe began as a point of concentrated potential, expanding and transforming into all that exists, a seed carries the blueprint for a tree's growth, demonstrating the universal principle of transformation.

Sociology
Seeds of thought and action propagate through human systems much like biological seeds in an ecosystem. They grow through collective energy into forces that reshape cultures and societies.

- *Example:* A revolutionary idea can spread and grow into a societal movement, mirroring the ripple effect of seeds dispersing and creating forests.

Psychology
A seed serves as a metaphor for the mind, where ideas germinate and grow when nurtured under the right conditions. Just as a seed develops into a tree, a single thought can take root, branch out, and create lasting change. This reflects the TBT principles of **cycles** and **transformation**.

- *Example:* A single inspiring idea—planted in the mind—can grow into a psychological "tree," influencing beliefs, behaviors, and even collective movements.

Economics
The concept of a seed mirrors the principle of **philanthropy**, where giving is the catalyst for growth and renewal. Just as a seed grows into something greater than itself, an act of generosity can create far-reaching benefits, aligning with the TBT principles of **reciprocity** and **transformation**.

- *Example:* A single act of financial support—such as funding education or community programs—plants the seed for opportunity, knowledge, and innovation. Over time, these benefits grow through society, creating cycles of progress and renewal.

The seed transcends its biological origins to illustrate the universal laws of TBT across disciplines, demonstrating how systems sustain themselves and evolve when aligned with natural principles.

Section 3: Observations and Universal Application
The seed is more than a biological entity—it is a universal model. Its transformation and influence ripple outward, shaping ecosystems, societies, and even the cosmos itself.

Biological Contagion
The growth of a seed ensures the continuation of life and the expansion of ecosystems.

> • *Example:* A single mangrove seed stabilizes soil, shelters marine life, and protects against erosion, demonstrating how one seed influences entire systems.

Social and Conceptual Contagion
Seeds of thought and action inspire further growth, creating cycles of change.

> *Example:* A single community garden project inspires collective action, fostering connection and shared resources.

The Universe as a Seed
Could the universe itself operate like a seed? The beginning echoes the germination of a seed, carrying within it the encoded potential for galaxies, stars, and life. The thought will expand on this idea later in the book.

> • *Example:* Just as a seed grows into a tree, the universe expands, governed by the same principles of **cycles**, **reciprocity**, and **transformation**.

The seed is not just a metaphor but a dynamic proof of the TBT's universality. It bridges disciplines, connects systems, and reveals the patterns that guide all existence, unveiling the underlying laws that shape everything we know.

Interdisciplinary Reflections in Summary: The Seed in Universal Context
A seed is not just a vessel of biological potential; it is a lens through which we observe the universal principles that govern all systems. Its growth transcends disciplines, making it a unifying symbol of the *Transcendent Blueprint Theory* **(TBT)**.

> • **Physics:**

The transformation of a seed mirrors the flow of energy in physics. Just as the universe began with concentrated potential that expanded into galaxies and life, a seed contains the blueprint for a tree. The cycles of energy and matter seen in **photosynthesis** reflect the same universal principles that drive the stars, planets, and cosmic systems.

- **Biology:**

A seed is the cornerstone of ecosystems, embodying cycles of **life, growth, and renewal**. As it develops, it interacts with its environment—exchanging energy, nutrients, and support—demonstrating **reciprocity** in nature. Its connections to pollinators, decomposers, and other organisms reveal the intricate web of **interdependence** that sustains all life.

- **Sociology:**

Seeds of thought and action propagate through human societies, much like biological seeds in a forest. A revolutionary idea or movement begins as a small spark but, through collective energy, grows into a force that reshapes cultures and systems. This ripple effect mirrors the dispersal and growth of seeds in nature.

- **Psychology:**

A single thought, like a seed, takes root in the fertile soil of the mind. With the right conditions—nourishment, repetition, and belief—it grows into habits, emotions, or ideologies. The spread of emotions or ideas within groups reflects **natural cycles of dispersal and growth**, showing how thought creates **influence** and transformation.

- **Economics:**

Investments, whether financial or social, are seeds of potential. They transform resources into opportunities, education, or innovation. The **cycles** of production, exchange, and renewal align with the principles of **reciprocity and transformation**, illustrating how economic systems evolve and sustain themselves.

The seed is more than a biological entity—it is a dynamic model of TBT's universality. It demonstrates that the principles of **cycles**, **reciprocity**, **transformation**, and **connection** transcend boundaries, shaping every system and discipline, from the physical to the conceptual.

Closing: Seeds as Proof of the Universal Blueprint

Every seed carries within it the potential for transformation, a living testament to the foundational principles of the *Transcendent Blueprint Theory* **(TBT)**. Encoded in its structure is not just the possibility of a tree, but the **cycles** of growth, the **reciprocity** of exchange, and the **renewal** that sustain life. Its transformation is both influenced by and influential to the systems it inhabits.

This understanding extends beyond biology. If a seed mirrors the principles of TBT, what does this reveal about the **universe itself**? Could the universe, like a seed, carry within its origins the blueprint for its expansion and transformation? The unfolding of galaxies, the formation of stars, and the evolution of life echo the growth of a seed into a tree—each governed by cycles, reciprocity, and transformation. This comparison invites us to consider not only the origins of existence but its trajectory and the patterns that guide it.

And what of humanity's role in this process? Every thought, action, or contribution we make **plants a seed**—whether in the soil of the earth, the hearts of others, or the systems we build. These seeds **grow and branch outward**, shaping the environments we inhabit and the futures we create. The proof of TBT is not abstract; it is tangible, alive in the seeds we plant and the systems they grow into.

The seed has revealed the encoded potential within the blueprint. Now we turn to the universal principles that guide this unfolding. In the next chapter, we will explore how the idea of the seed extends to all domains of existence, formalizing the **Seed Law** as a universal truth that governs transformation and growth across all disciplines.

Chapter 3.2: The Seed Law

Introduction: The Seed Law as a Universal Truth

At the heart of the *Transcendent Blueprint Theory* (TBT) lies one of many foundational discoveries: **The Seed Law**. This scientific principle posits that the potential of any system—whether physical, biological, or conceptual—is encoded within its origin. Like a seed holding the blueprint for an entire tree, every system carries within it the latent potential for growth, transformation, and impact.

The Seed Law does not merely describe a biological phenomenon. It transcends disciplines, revealing how universal principles such as cycles, reciprocity, and transformation govern the unfolding of potential across all forms of existence. From the physical expansion of the universe to the intellectual evolution of ideas, the Seed Law offers a unifying framework for understanding how systems progress from origin to expression.

The Seed Law: Encoded Potential Across Systems
Definition:

The Seed Law states that the potential of any system is encoded at its origin and unfolds through interaction with external forces according to universal principles such as cycles, transformation, and connection. This law applies universally, encompassing physical, biological, and conceptual systems.

1. Physics: The Seed of the Universe

Encoded within the beginning was the potential for galaxies, stars, planets, and life itself. Like a seed, the singularity carried the blueprint for an expanding universe shaped by cycles, transformation, and interconnection.

- *Example:* The gravitational forces and energy flows established during the universe's inception acted as the roots, anchoring the cosmos in its trajectory toward complexity. The expansion of space mirrors the growth of a seed into a tree, with each stage governed by universal laws.
- *TBT Insight:* The universe itself demonstrates the Seed Law, as its initial conditions contained the encoded potential for all subsequent structures and systems. This unfolding reflects the same principles observed in biology and thought: cycles of

expansion and contraction, transformation of energy into matter, and the connection between all components.

2. Biology: Seeds as the Blueprint for Life

A seed is nature's purest representation of encoded potential. Within its protective shell lies the DNA blueprint for growth, reproduction, and adaptation. However, its transformation depends on external catalysts such as sunlight, water, and nutrients.

- *Example:* The life cycle of an oak tree begins with an acorn. This tiny vessel holds the latent potential for roots to stabilize soil, branches to provide shelter, and leaves to produce oxygen. Each stage of growth reflects cycles of reciprocity with its environment, sustaining ecosystems in return.
  - *TBT Insight:* The seed's journey mirrors universal principles: cycles of dormancy and growth, transformation of energy into matter, and connection to its environment. The seed depends on external forces to activate its growth, just as it contributes back to the system that sustains it.

3. Thought: Seeds of Ideas

Ideas, much like seeds, carry the blueprint for innovation and transformation. Encoded within each idea is the potential to grow outward, influencing individuals, communities, and entire cultures.

- *Example:* The discovery of electricity began as a seed of thought. Through **resistance** (skepticism, experimentation) and **connection** (collaboration, application), this idea grew into a force that reshaped societies and transformed human life.
- *TBT Insight:* The propagation of ideas follows the same principles as biological seeds. Just as a tree disperses seeds to perpetuate its species, ideas spread through cycles of discussion, refinement, and resistance, reshaping the systems they inhabit. However, **echo chambers**—where the same ideas are repeated and reinforced—act like **seed vaults**, preserving and propagating only a single "species" of thought. In contrast to diverse ecosystems, this limits growth, transformation, and innovation.

Universal Principles in Action

The Seed Law reveals a universal truth: potential is encoded at the origin of every system and unfolds through interaction with external forces.

1. **Cycles: Seeds are part of ongoing cycles of growth, dormancy, and renewal, reflecting the rhythms that sustain all systems.**
2. **Reciprocity: The growth of a seed depends on its environment, just as its fruition contributes back to the system that nurtured it.**
3. **Transformation: A seed's potential is realized through change, turning dormant energy into dynamic form.**
4. **Connection: No seed grows in isolation; its development depends on and contributes to the systems around it.**
5. **Motion: A seed activates and expresses its potential through movement—expanding roots, reaching upward, and dispersing its next generation—demonstrating the dynamic nature of growth.**

Implications of the Seed Law

The Seed Law has profound implications across disciplines:

- **In Physics**: It frames the universe as a dynamic system whose growth mirrors the principles of cycles, transformation, and connection.
- **In Biology**: It highlights the interdependence of organisms and ecosystems, emphasizing the role of reciprocity in sustaining life.
  - **In Thought and Sociology**: It reveals how ideas and actions propagate, transforming societies through cycles of resistance and refinement.

By understanding the Seed Law, we gain a lens through which to view all systems, from the smallest seed to the vast expanse of the cosmos.

Conclusion: A Foundation for Universal Understanding

The **Seed Law** is not merely a concept; it is a universal truth that bridges disciplines and unites systems under the framework of the **Transcendent Blueprint Theory (TBT)**. It demonstrates how **encoded**

potential drives growth, transformation, and renewal, shaping the trajectories of physical, biological, and conceptual systems alike.

As we move forward, we will explore other foundational principles of TBT, revealing how the universal truths observed in the seed extend to **every facet of existence**. The Seed Law stands as a testament to the power of potential—a reminder that within every beginning lies the blueprint for infinite possibilities.

This book itself is a seed—an idea rooted in universal truths, planted with the intent to grow into new ways of thinking, acting, and understanding the world. Its potential lies in the hands of those who nurture it, allowing its principles to take root, branch outward, and transform systems across disciplines and lives.

Part 2
What Are the Rules

Chapter 4: Foundations of Universal Laws

Introduction: Unveiling the Rules
Imagine the cycles of the seasons, the ripple of an idea spreading through a crowd, or the perpetual motion of celestial bodies. Beneath each of these phenomena lies a shared set of universal principles—rules that transcend boundaries, shaping both the tangible and the intangible. These foundational truths guide existence itself, connecting the vastness of the cosmos to the smallest spark of thought.

Every system, from the simplest particle to the most complex ecosystem, operates under a set of principles—universal laws that govern and sustain existence. These laws are not confined to one domain; they transcend disciplines, appearing in physics, biology, thought, and society.

The *Transcendent Blueprint Theory* **(TBT)** does not replace these established laws. Instead, it reveals them, showing how they align under shared principles like cycles, reciprocity, transformation, motion and resistance, and connection. Through the lens of TBT, these laws come alive as more than mere observations; they become the framework through which all systems operate and interact.

Before we can explore how TBT interacts with specific laws like Newton's laws of motion or the conservation of energy, we must first lay the foundation: What are these universal principles, and how do they shape our understanding?

Universal Principles: The Framework of Connection
At first glance, the laws governing physics, biology, and thought may seem distinct. However, through the lens of the *Transcendent Blueprint Theory* (TBT), a deeper pattern emerges, revealing that these laws are expressions of universal principles. These principles are not isolated to one domain but transcend disciplines, guiding both the physical and intangible realms.

Beneath every observable law lies a universal principle that shapes existence itself. These five foundational principles—cycles, reciprocity, transformation, motion and resistance, and connection—serve as the scaffolding that underpins existence. They govern the cycles of planets, the course of thought, and the structure of human societies, showing us how everything, from the smallest particle to the largest cosmic body, operates under these forces.

1. Cycles: The Rhythm of Continuity

Life unfolds in cycles: the turning of seasons, the orbit of planets, and the repetitive patterns of thought and behavior. These rhythms ensure continuity and renewal, weaving a dependable structure into the chaos of existence.

- *Example:* Newton's First Law of Motion (inertia) embodies this principle in the physical world, describing the perpetual motion of objects unless interrupted by an external force.
- **Biological Insight**: Similarly, the cycles of birth, growth, death, and regeneration reflect how decay fuels renewal. The fall of one process seeds the beginning of another, mirroring patterns found in forests, where fallen trees enrich the soil for new growth.
- **Cosmic Parallel**: The orbits of celestial bodies, from moons to galaxies, demonstrate cycles on an incomprehensible scale, connecting the smallest rotations to the grandest motions.

In human societies, cycles are evident in the rise and fall of civilizations. Each era brings renewal, building on the remnants of the past. The pendulum of time swings through growth, stagnation, and transformation, reflecting nature's own rhythms.

Cycles ensure systems endure, and change, through moments of growth, rest, and renewal across time.

2. Reciprocity: The Balance of Exchange

In the interplay of energy and resources, reciprocity sustains balance. It is the give-and-take that fosters harmony within ecosystems, economies, and even interpersonal relationships.

- *Example:* Newton's Third Law (action and reaction) reflects reciprocity, where every force generates an equal and opposite response.
- **Biological Insight**: Symbiotic relationships, like the exchange between bees and flowering plants, demonstrate this principle in ecosystems. Without this reciprocity, life collapses into imbalance.

- **Human Connection**: Economic systems thrive when reciprocity underpins exchanges. A transaction, whether financial or relational, reflects the dynamic balance between giving and receiving.

Even on the molecular level, this balance persists. Within the body, homeostasis depends on the reciprocal flow of nutrients, oxygen, and waste. The intricate dance of these exchanges mirrors the equilibrium that sustains the universe.

Reciprocity connects systems through the continuous flow of energy, ensuring no element exists in isolation.

3. Transformation: The Power of Change

Transformation is the process by which energy and matter evolve into new forms, enabling growth, adaptation, and renewal. It is the force that turns potential into reality.

- *Example:* The Second Law of Thermodynamics illustrates transformation, where energy dissipates and changes form.
- **Biological Insight**: Photosynthesis transforms sunlight into chemical energy, fueling the growth of plants and sustaining entire ecosystems.
- **Sociological View**: The transformative power of collective movements reshapes societies. Revolutions and innovations are moments of transformation, born from challenges and potential.

Transformation is also evident in the life of a seed. From dormancy to growth, it undergoes profound changes, exemplifying nature's capacity for renewal. Similarly, human thought transforms through learning, adapting to new information and perspectives.

Transformation drives innovation and adaptation, revealing how systems respond to their surroundings.

4. Motion and Resistance: The Refinement of Progress

Movement propels systems forward, while resistance hones and stabilizes that progress. This interplay creates the friction necessary for refinement and growth.

- *Example:* In physics, friction shapes motion, ensuring control and stability.
- **Societal Insight**: Social movements gain strength through resistance. The friction of opposition refines purpose, strengthening collective resolve.
- **Thought Process**: Ideas mature under scrutiny. The resistance they encounter refines their structure, ensuring that only the most resilient and impactful thoughts take root.

Even in ecosystems, resistance shapes adaptation. Predators and prey change together, their interactions refining each other. This balance ensures the survival of both, maintaining the intricate web of life.

Resistance is not an obstacle but a catalyst for growth, challenging systems to become stronger and more resilient.

5. Connection: The Web of Interdependence

No system exists in isolation. Connection binds all elements of existence, creating networks of interdependence that sustain life and structure the universe.

- *Example:* Gravity connects celestial bodies into cohesive systems, while ecosystems rely on the interdependence of species.
- **Social Insight**: Human societies thrive when connections are fostered, emphasizing collaboration over isolation.
- **Cognitive Perspective**: Ideas and relationships grow through networks. The more connections they foster, the stronger and more enduring their impact.

From the mycorrhizal networks of forests to the neural pathways of the brain, connection is the foundation of complexity. It is through these links that systems gain resilience and adaptability.

Connection reveals the unity beneath diversity, showing how systems rely on one another to thrive and adapt.

Bridging Established Laws with the Transcendent Blueprint Theory

For centuries, scientific laws like Newton's Laws of Motion, the Laws of Thermodynamics, and Darwin's Theory of Evolution have provided humanity with tools to attempt to decode the complexities of the universe. Each law illuminates specific phenomena within its domain, but what if they are only fragments or observations of a larger framework?

The *Transcendent Blueprint Theory* (TBT) doesn't aim to replace all foundational laws but to complement, clarify or unify them. (some will be challenged like Darwin's Evolution, or our thoughts about Magnetism) By framing them as components of a shared blueprint, the TBT reveals the deeper principles—cycles, reciprocity, transformation, motion and resistance, and connection—that govern existence across all disciplines.

Physics: The Mechanics of Existence

Physics provides the foundational framework for understanding how the universe operates. It reveals the interplay between motion, energy, and forces, offering insights into universal principles such as cycles, resistance, and transformation. Through the lens of the *Transcendent Blueprint Theory* (TBT), we see how these principles manifest in the physical world and extend into other domains.

Newton's Laws of Motion: Cycles, Resistance, and Friction

Newton's laws of motion reveal fundamental truths about movement, stability, and the forces that govern change. Each law aligns with universal principles like cycles, reciprocity, and transformation, showcasing how systems refine themselves through interaction with resistance and motion.

Friction as Stability

Friction, often seen as a hindrance, is a stabilizing force that transforms chaotic motion into purposeful action. Without friction, motion would lack control, leading to disorder. Instead, friction serves as the crucible for refinement, enabling progress.

- *Example:* A wheel rolling down a slope depends on friction to maintain grip and ensure forward progress. Without it, the motion would spiral into chaos, losing direction and stability.
- *TBT Insight:* Friction exemplifies resistance as a force of growth. It mirrors how intellectual debates refine ideas, societal challenges catalyze change, and biological adaptation strengthens organisms. Resistance transforms

unstructured motion into stability, demonstrating its universality across systems.

Newton's First Law: Cycles and Resistance in Action

The law of inertia states that objects in motion remain in motion unless acted upon by an external force. This principle reflects the cyclic rhythms inherent in systems—steady motion until resistance initiates transformation.

- **TBT Perspective:** Resistance does not halt progress; it refines it. This interplay between motion and resistance ensures systems transcend stagnation, achieving renewed functionality. Across disciplines, cycles of motion and renewal reveal how external forces drive refinement and growth.

Newton's Third Law: Reciprocity and Transcendence

The law of action and reaction illustrates reciprocity: for every action, there is an equal and opposite reaction. This balance maintains motion and harmony, ensuring that systems remain interconnected.

- **TBT Perspective:** Reciprocity is not merely about balance but about transcendent exchange. It bridges systems, fostering interdependence and renewal. Whether in the physical motion of objects or the exchange of energy within ecosystems, reciprocity reveals a transformative interplay that transcends boundaries, enabling growth and sustained progress.

Conservation of Energy: The Transformation of Potential

The First Law of Thermodynamics states that energy cannot be created or destroyed—it only changes form. This fundamental truth reflects the TBT principle of transformation, where systems evolve by converting potential energy into action and renewal.

- *Example:* Photosynthesis transforms sunlight into chemical energy, fueling growth and sustaining life. Similarly, in physics, energy flows and shifts forms, driving cycles of renewal across the cosmos.
- **TBT Perspective**: Transformation governs the flow of energy across all systems, transcending physical phenomena. Energy exists not only in the physical realm but also in the intangible

dimensions of human thought, relationships, and ideologies. The potential of an idea can ignite social movements, just as emotional energy exchanged in relationships catalyzes growth. Transformation reveals how systems adapt, persist, and change through cycles of renewal, uniting tangible and intangible processes under one universal principle.

Connection: The Fabric of the Universe

Connection is the thread that binds all elements of the physical universe, creating networks of interdependence that sustain and shape existence. Forces like gravity exemplify how systems depend on connection to maintain cohesion and structure.

- *Example:* Gravity links celestial bodies into cohesive systems, balancing their motion and ensuring stability across the cosmos. It is the invisible web that holds planetary systems together, providing a foundation for harmony and balance.
- **TBT Perspective**: Connection reveals the unity beneath diversity, illustrating how systems across the universe depend on interdependence to thrive. Whether through gravitational forces binding galaxies or the transfer of energy between ecosystems, connection ensures that no system exists in isolation. It extends beyond the physical, manifesting in the shared experiences of human communities and the mutual support that binds societies together. Connection, as a universal principle, bridges the divide between the tangible and intangible, linking all forms of existence into a cohesive whole.

Cycles: The Rhythm of Continuity

Physics illuminates the cyclical nature of existence. From the revolutions of planets to the oscillations of energy, cycles provide the rhythm that ensures continuity and renewal.

- *Example:* Newton's First Law (inertia) illustrates cycles in motion, where objects persist until external forces create new cycles of action. The orbits of planets around stars and the repetitive nature of energy oscillations reflect the rhythmic structure embedded in the universe.
- **TBT Perspective**: Cycles are the foundation of stability and progress, weaving moments of growth, rest, and renewal into

the fabric of all systems. These rhythms not only maintain the integrity of physical systems but also mirror the cycles of life, thought, and human endeavor. By integrating rest and renewal with action and growth, cycles ensure the resilience and adaptability of systems, revealing their universal applicability across disciplines.

Transitioning to Biology: The Blueprint of Life

Physics lays the groundwork for understanding universal principles—motion, resistance, energy transformation, and connection—that underpin all systems. These same principles resonate in biology, where cycles, reciprocity, and connection shape life's intricate web of relationships and processes. As we move into the realm of living systems, we will uncover how the universal laws of physics inform and guide the dynamic processes of organisms, ecosystems, and adaptation.

From Physics to Biology: Energy and Adaptation

The principles governing physical systems often find striking parallels in biological processes, revealing a shared foundation across disciplines.

Conservation of Energy: Cycles and Transformation

In physics, energy is neither created nor destroyed; it cycles through systems, transforming from one form to another.

- **Biological Parallel**: In biology, this mirrors nutrient and energy exchanges within ecosystems, where energy flows from sunlight to plants, from herbivores to predators, and finally back into the soil through decomposition.

- *TBT Insight:* The cyclical transformation of energy underscores the universality of conservation and renewal, uniting physics and biology under the principle of **transformation**. It demonstrates how systems sustain themselves and adapt through energy exchange and conversion.

Motion and Resistance: Refinement Through Challenge

In physics, motion is refined by resistance, such as friction shaping controlled movement.

- **Biological Parallel**: Similarly, biological systems change through the resistance imposed by environmental pressures, driving adaptation and fostering resilience.
- *TBT Insight:* The interplay of motion and resistance highlights how challenges act as catalysts for growth and refinement, whether in the development of physical systems or the adaptation of living organisms to their environments.

Biology: The Blueprint of Life

Life is governed by universal principles that manifest through intricate biological processes. These principles, observable in **adaptation**, **symbiosis**, and the pressures exerted by the environment, illustrate how living systems align with the foundational truths of the *Transcendent Blueprint Theory* (TBT).

Adaptation: Transformation in Motion

Adaptation is the dynamic process through which species respond to environmental challenges, reflecting the TBT principles of transformation, motion, and connection.

- *Example:* The finches of the Galápagos Islands have adapted beak shapes to match the food sources available in their environments. These changes, which are said to be observable within short timescales, may highlight how environmental pressures drive transformation and refinement in living systems.
- *TBT Insight:* Adaptation reveals a process of cyclical transformation. It aligns species with their surroundings through **motion**, as they refine their traits, and through **connection**, as they integrate into symbiotic relationships and ecosystems. This principle also echoes in societal systems, where change and progress emerge as responses to challenges, ensuring continuity and renewal.

Symbiosis: Reciprocity in Action

Symbiosis exemplifies reciprocity, as organisms exchange resources and benefits to sustain balance and interdependence. From mutualistic relationships between plants and pollinators to the microscopic partnerships of fungi and roots, symbiosis highlights the interconnected nature of life.

- *Example:* Mycorrhizal fungi form networks with plant roots, exchanging nutrients for carbon, enhancing soil health and plant vitality.
- *TBT Insight:* Symbiosis in biology parallels human relationships and societal structures, illustrating how cooperation fosters resilience and mutual growth. This reciprocity underscores the transcendent nature of interdependence, which is vital for both ecological and social stability.

Biology: Pressure Shapes Adaptation

Environmental pressures act as the crucibles of biological adaptation. These challenges compel organisms to create traits that ensure survival and reproduction, demonstrating the principle of **resistance** as a force for refinement.

- *Example:* Plants in arid climates develop root systems to access water, while desert animals develop behaviors and structures to conserve moisture.
- *TBT Insight:* Adaptation through resistance exemplifies the transformative power of challenges. This principle echoes across systems, from the strategies of organisms to the intellectual and societal progress driven by adversity. Resistance serves as a universal mechanism that refines systems, enabling them to thrive in changing environments.

Society and Thought: The Intangible Blueprint

The principles that govern physical and biological systems extend seamlessly into the intangible realms of thought and society. These universal truths—**cycles**, **transformation**, **reciprocity**, **motion**, **resistance**, and **connection**—shape the evolution of ideas, the structure of human relationships, and the dynamics of societal progress. Through the lens of the *Transcendent Blueprint Theory* (TBT), we uncover how thought and sociology mirror the same natural laws that govern ecosystems and galaxies.

The Ripple Effects of Ideas: Cycles in Thought

Ideas, like waves in water, propagate outward, influencing and reshaping the collective consciousness. This cyclical spread mirrors the movement of energy in physics or the dispersal of seeds in biology.

- *Example:* A revolutionary idea, such as the abolition of slavery, begins as a small ripple, gradually influencing larger societal movements. Over time, it transforms collective beliefs and spurs action.

- *TBT Insight:* The spread of ideas reflects cycles of **motion** and **transformation**. Each idea acts as a seed, growing through cycles of resistance, refinement, and acceptance. This principle demonstrates the universality of growth, connecting thought to the broader patterns of life and the cosmos.

Reciprocity in Relationships: The Balance of Exchange

Human relationships thrive on reciprocity, where the exchange of support, energy, and resources sustains balance and fosters connection. This dynamic parallels the symbiosis found in biological systems, where organisms mutually benefit from shared exchanges.

- *Example:* Acts of kindness and generosity create a ripple effect, fostering trust and strengthening communal bonds. These exchanges build a foundation for societal stability and growth.

- *TBT Insight:* Reciprocity bridges the tangible and intangible, highlighting how systems—whether social or ecological—depend on balanced exchanges to thrive. This principle reveals that **connection** is the thread that binds diverse systems, ensuring cohesion and harmony.

Resistance: The Catalyst for Progress
Thought: Opposition Strengthens Ideas

Ideas, much like ecosystems, grow stronger under scrutiny. Resistance serves as a refining force, challenging weaknesses and fostering growth.

- *Example:* The scientific method thrives on critique and debate, with each cycle of resistance strengthening the validity of hypotheses and theories.

- *TBT Insight:* Resistance embodies the principle of **motion** and **transformation**, ensuring that only the most resilient ideas endure. This interplay aligns with the cycles observed in ecosystems and the refinement of energy in physical systems.

Sociology: Resistance Fuels Collective Change

Societal change often emerges from resistance. Movements gain strength and clarity when confronted by opposition, galvanizing communities to achieve shared goals.

- *Example:* The civil rights movement in the United States, spurred by systemic resistance, transformed societal norms and legal frameworks, advancing equality.
- *TBT Insight:* Resistance in society mirrors the forces that shape ecosystems, where environmental pressures refine adaptation. By aligning with cycles of challenge and renewal, resistance ensures the continual evolution of human systems.

Universal Principles in Thought and Society

The universal principles of the TBT—**cycles, reciprocity, transformation, motion and resistance,** and **connection**—transcend the physical and biological, shaping the intangible realms of thought and society.

- **Cycles**: Ideas and social movements follow rhythms of growth, resistance, and renewal, echoing the patterns of natural systems.
- **Reciprocity**: Balanced exchanges of energy, resources, or support sustain relationships, communities, and systems.
- **Transformation**: Change emerges through cycles of challenge and adaptation, ensuring progress, resilience, and sustainability.
- **Motion and Resistance**: Systems remain dynamic and responsive, refining themselves through movement and interaction. Resistance acts as a force of refinement, ensuring that challenges drive systems toward greater alignment, growth, and resilience.
- **Connection**: The thread that binds systems into cohesive, thriving networks across disciplines, ensuring integration and interdependence.

Conclusion: Toward a Unified Blueprint

The Transcendent Blueprint Theory reveals that universal laws are not isolated silos of knowledge. Instead, they are expressions of shared principles—guiding forces that shape every aspect of existence. From the vast expansion of galaxies to the delicate germination of seeds, from the relentless motion of particles to the quiet ripple effects of ideas, these laws illuminate the universe as a tapestry of unity.

Through the lens of TBT, we see that life and the cosmos are bound by cycles of repetition and renewal, transformed through resistance, and sustained by reciprocity and connection. The same principles that guide the growth of a tree into a forest also influence the change of societies, the flow of energy through ecosystems, and the emergence of thought into action.

But these laws do more than unify—they inspire us to look deeper, beyond the surface of isolated systems, to uncover the shared rhythms that govern all existence. They challenge us to see that progress is not linear, but cyclical; that resistance is not an obstacle, but a tool for refinement; and that connection is not just an advantage, but a fundamental necessity for survival and growth.

As we move forward, this book will follow the unfolding of these universal principles across disciplines, revealing their profound impact on our understanding of biology, physics, thought, society and beyond. By exploring these foundational truths through the TBT, we begin to uncover not just how systems function, but how they thrive in harmony with the blueprint that unites them all.

The journey ahead is both scientific and philosophical—a path to seeing the world, and ourselves, through the universal lens of connection, transformation, and reciprocity. Together, let us take the next step toward understanding the principles that sustain existence, guiding both what is and what could be.

Chapter 5: Biology — The Cycles of Life

Introduction: The Language of Life

Life operates through universal patterns, governed by **cycles, reciprocal relationships, transformation, motion and resistance,** and **connection**. From the smallest cell to the vast complexity of ecosystems, these principles connect biology to broader systems like thought, society, and even the cosmos, as seen through the lens of the *Transcendent Blueprint Theory* (TBT).

Consider a forest: fallen leaves decompose into the soil, nurturing new growth. Water flows through roots, rises into the atmosphere, and returns as rain. These processes not only sustain life but reveal universal truths about balance, renewal, and the universality of existence. What can biology teach us about resilience and sustainability across all systems?

1. The Cycles of Life: Nature's Perpetual Motion

Biology thrives on **cycles**—predictable rhythms that ensure sustainability and renewal. These natural patterns highlight the fundamental law that nothing is wasted, and every process feeds into the next.

- **Nutrient and Water Cycles**: Carbon, nitrogen, and water cycle through organisms and ecosystems, sustaining life through renewal and transformation. For example, plants fix carbon from the atmosphere, which animals consume and return to the soil through decomposition.
- **TBT Parallel in Thought**: Just as nutrients cycle through ecosystems, thoughts evolve through cycles of inspiration, refinement, and memory, providing fertile ground for new ideas.
- **Ecosystems as Circular Systems**: The death of one organism fuels the growth of others, mirroring the cyclical nature of economic and societal systems.
- **TBT Parallel in Humanity**: Economic booms and recessions mirror biological cycles of growth and dormancy. Systems thrive when they align with these natural rhythms.

Biological cycles reflect universal rhythms. Across ecosystems and economies, repetition ensures sustainability by fostering adaptation and renewal.

2. Reciprocity: The Gift Economy of Nature

Reciprocity is nature's guiding principle, where every organism contributes as much as it takes to maintain balance. This exchange is not a one-sided transaction but a mutual relationship that sustains systems across time.

- **Biological Reciprocity**: Mycorrhizal fungi form vast underground networks, exchanging nutrients with trees in a mutually beneficial relationship. Similarly, predator-prey dynamics maintain ecological balance, ensuring that no single species dominates the system.
- **TBT Parallel in Humanity**: Acts of generosity, like donations or mutual aid, mirror the reciprocity found in ecosystems. These exchanges foster harmony and renewal in human communities, much like nature's systems thrive on mutual contribution.
- **Symbiotic Balance**: Plants absorb carbon dioxide and release oxygen, forming an interdependent relationship with animals that ensures both species thrive.
- **TBT Parallel in Thought**: Just as ecosystems maintain balance through reciprocal exchanges, ideas thrive in an environment of dialogue and feedback. Thought, when shared and challenged, refines itself, creating intellectual ecosystems that mirror nature's balance.

Reciprocity is the foundation of balance and sustainability. From biological systems to human thought and interaction, giving and receiving sustain the cycles of growth, renewal, and harmony.

3. Transformation: Adaptation and Renewal

Transformation is the essence of survival. Biological systems adapt, aligning themselves with external forces to persist and thrive.

- **Adaptation as Transformation**: Species make changes to fit their environments, exemplifying resilience through change. For example, plants in arid climates develop roots to access water.
- **TBT Parallel in Humanity**: Social and cultural adaptation mirrors biological transformation. Societies evolve structures to meet challenges, much like species adapt to survive.

- **Keystone Species**: Certain organisms, like coral reefs or wolves, drive transformation within ecosystems, reshaping the balance of entire systems.
- **TBT Parallel in Thought**: Keystone ideas—revolutionary concepts like this one—can reshape disciplines or societies, much like keystone species transform ecosystems.

Transformation demonstrates the resilience of systems. By embracing change, both natural and human systems align with universal principles of renewal and progress.

4. Connection: The Web of Life

Connection is the principle that binds biological systems into a cohesive whole. Each organism contributes to and depends on its environment, creating networks of mutual reliance and resilience.

- **Biodiversity as Resilience**: Ecosystems thrive when they are diverse. The loss of one species can often be compensated for by others, preserving the system's overall balance and adaptability.
- **TBT Parallel in Thought**: Diversity of ideas strengthens societal problem-solving, allowing groups to adapt to complexity. Monocultures—whether in ecosystems or thought—create fragility and hinder growth.
- **The Ripple Effect**: When pollinators like bees decline, the effects cascade through ecosystems, disrupting food chains and threatening biodiversity.
- **TBT Parallel in Humanity**: The collapse of a financial institution sends shockwaves through connected industries, demonstrating how fragile systems suffer when over-reliance replaces balanced connection.

Connection fosters resilience and adaptability. The networks observed in biological systems mirror the reliance and contributions required in human societies and thought. By recognizing these connections, systems can align with principles that sustain growth and recovery.

5. Motion and Resistance: Forces of Refinement

In biological systems, **motion** is the active response to challenges, while **resistance** ensures refinement and alignment. Together, they drive adaptation and progress.

- **Biological Motion and Resistance**: Roots move through the soil in search of water, while environmental resistance—such as drought—pushes plants to refine their structures. Predator-prey dynamics refine species over time through constant interaction.
- **TBT Parallel in Thought**: Intellectual motion occurs as ideas spread, adapt, and evolve. Resistance, such as critique or opposition, refines ideas, ensuring their resilience and alignment with truth.
- **Stress and Adaptation**: Trees growing in windy conditions develop stronger wood fibers as a response to resistance. This motion-resistance interplay creates resilience in biological systems.
- **TBT Parallel in Humanity**: Societies, like ecosystems, grow stronger when challenged. Resistance refines systems, ensuring progress is sustainable and aligned with universal principles.

Motion and resistance reveal that systems—whether biological, societal, or intellectual—thrive through challenge. Motion enables growth, while resistance ensures refinement and resilience.

Biology serves as a profound teacher, demonstrating how **cycles**, **reciprocity**, **transformation**, **motion and resistance**, and **connection** guide the growth and resilience of living systems. These principles, far from being confined to the natural world, transcend disciplines—offering insights that unify physics, thought, society, and the broader cosmos.

Conclusion: Life as a Universal Blueprint

Biology is not merely the study of life; it is the study of the universal principles that sustain all systems. Through cycles, reciprocity, transformation, and connection, biology provides a model for resilience and renewal that transcends ecosystems and disciplines.

The lessons of biology remind us that thriving systems align with universal laws. By observing the patterns of life—whether in nurturing ideas, fostering resilience, or adapting to challenges—we can apply these principles to thought, society, and beyond. As we move forward, we will explore how these same principles guide physics, human systems, and the intangible realms of thought, revealing the unified framework that connects all disciplines.

Biology serves as the first thread in this tapestry, linking the natural world to the universal truths of existence. It reminds us that the forces shaping life are the same forces that govern the cosmos, guiding us toward understanding, growth, and harmony with the blueprint of existence.

Chapter 6: Physics — The Mechanics of Existence

Introduction: Physics as a Universal Key

Physics is the language of existence. It describes how matter and energy interact, shaping the universe from the smallest particles to the largest galaxies. But when viewed through the lens of the *Transcendent Blueprint Theory* (TBT), physics becomes more than a study of motion, force, and energy—it becomes a framework for understanding how systems interact, adapt, and transform across disciplines.

This chapter explores the core principles of physics—motion, resistance, transformation, and connection—and bridges them to biology, thought, and society. The universality of these principles reveals how the same rules that govern the cosmos shape our daily lives, relationships, and even the flow of ideas.

1. Motion and Inertia: The Power of Momentum

Newton's First Law states that an object in motion stays in motion unless acted upon by an external force. This principle, often referred to as inertia, applies not only to physical objects but also to systems, habits, and even lives.

- **Physical Inertia:**

A rolling ball continues to move until friction or another force slows it down.

TBT Parallel: In human behavior and ecosystems, both forward momentum and stagnation have inertia. For instance, once a system begins to decline, the forces driving that decline can compound, creating a spiral that is difficult to reverse.

- **Inertia and the Decline Spiral:**

In biological systems, a tree experiencing stress often enters a decline spiral. Once weakened, the tree becomes more susceptible to additional stresses—pests, disease, or drought—that accelerate its decline.

Example: A stressed tree may lose its leaves early, reducing its ability to photosynthesize. This lack of energy compounds its stress, making recovery less likely without significant external intervention.

Human Parallel: In personal or organizational contexts, stagnation or negative momentum often leads to compounding challenges. For example, a business facing declining sales may cut corners, leading to reduced quality and further loss of customers—a self-perpetuating cycle.

- **Reversing the Spiral:**

Breaking inertia in a decline spiral requires an external force to redirect momentum. In ecosystems, this could involve removing stressors or providing targeted resources. In human systems, it might mean introducing new energy, ideas, or strategies to reverse negative trends.

Example: For a tree, deep watering or nutrient supplementation can interrupt the decline spiral, giving it the resources needed to recover. Similarly, for a struggling team, a fresh perspective or new leadership can provide the spark needed to restore positive motion.

Inertia doesn't only sustain forward progress—it can also perpetuate decline. Recognizing and addressing negative inertia is essential to redirecting systems toward growth and balance.

2. Magnetism and the Dynamo Effect: The Push and Pull of Motion

Magnetism begins with motion. Within the Earth's core, charged particles circulate, generating the magnetic field that shields the planet and aligns its systems. This is the dynamo effect—a force born from motion, quietly shaping our world and sustaining life. Yet magnetism isn't confined to the realm of physics; it echoes throughout human systems, illustrating the profound interconnectedness of action, influence, and resistance.

The Physics of Magnetism

At its foundation, the current definition of magnetism (which we will challenge later) arises from the motion of particles. This movement creates invisible forces that attract or repel objects, defining their interactions. The Earth's magnetic field is a system in balance, its dynamism stabilizing the planet's environment.

- *TBT Insight:* In physics, motion creates influence. Magnetic fields extend far beyond their origin, shaping interactions across systems. Similarly, human actions generate invisible fields of influence that ripple through relationships, communities, and organizations.

A Tangible Illustration: Magnetism in Action

Consider a child dragging a piece of metal across cement. As the metal scrapes the surface, the friction creates heat and energy, dislodging tiny particles. Some of these particles adhere to the metal, while others are repelled, scattering across the ground. This simple act reveals the dual nature of magnetism: attraction and repulsion, shaped by motion and resistance.

- *Example:* The particles clinging to the metal illustrate magnetism's ability to create new connections through interaction. Similarly, the scattering of particles mirrors how friction and resistance refine systems by separating incompatible elements.

The Parallel in Human Systems

In human life, motion creates more than momentum—it generates magnetism. A single action, whether deliberate or reflexive, draws others closer or pushes them away. In crowded spaces, hurried movements can create friction, disrupting harmony in subtle but tangible ways.

- *Example:* Imagine a person rushing through a narrow kitchen while another prepares a meal. The swift motion creates tension, as though the hurried movement exerts an invisible push. This mirrors the way unaligned magnetic fields repel each other.

But when motion is deliberate and collaborative, it draws others in, creating harmony and shared purpose. A leader's decisiveness, much like a magnetic pole, can align a team, focusing their energy toward a common goal.

- *Example:* Consider a politician whose confident steps and purposeful motion inspire supporters. Their energy becomes magnetic, pulling others toward their vision. Conversely, overly aggressive or erratic actions can repel, creating opposition instead of unity.

Magnetism in Relationships

Magnetism governs relationships as well. In physics, aligned magnetic fields strengthen one another, creating stability and cohesion. Similarly, aligned actions and goals amplify human connection, fostering trust and synergy. When goals diverge or actions conflict, the result is resistance—a repelling force that strains bonds.

- *TBT Insight:* The alignment of motion is key. Just as misaligned magnetic fields create turbulence, conflicting actions or misaligned intentions in human systems lead to friction and diminished progress.
- *Example:* A high-performing team functions like a coherent magnetic field, where shared goals and mutual trust amplify results. By contrast, a team plagued by misaligned objectives experiences resistance, much like opposing poles of a magnet.

The Interplay of Motion and Resistance

The dynamo effect illustrates that motion alone is not enough—resistance plays an equally vital role. Without the friction generated by the Earth's molten core, its magnetic field would not exist. Similarly, human systems require opposition and tension to refine ideas, strengthen relationships, and ensure progress.

- *TBT Insight:* Motion and resistance are partners in growth. In the absence of resistance, motion lacks direction and purpose. Friction refines, stabilizes, and sharpens the systems it encounters.

Magnetism Beyond Physics

The parallels between magnetism in physics and human systems reinforce the universality of TBT principles. The push and pull of motion in our lives mirrors the interplay of attraction and repulsion in the physical world, connecting the tangible and intangible.

- **Inertia and Decline:** When motion stagnates, systems begin to spiral downward. This decline spiral reflects the importance of continuous movement and intentional direction, as inertia alone cannot sustain progress indefinitely.
- **Human Magnetism:** Whether through words, actions, or decisions, our motions generate influence. A politician's campaign, a parent's guidance, or an artist's creativity creates ripples of attraction and resistance that shape their surroundings.

Magnetism in physics is a profound illustration of motion's transformative power, but its true universality lies in its parallels across disciplines. Whether stabilizing planetary systems, building human connections, or refining ideas through resistance, the underlying principle of magnetism connects all systems. By understanding its push and pull, we can align our actions to create influence, minimize unnecessary friction, and foster cohesive systems of growth and purpose.

The Dynamo of Thought: Motion Sparks Innovation

Imagine a single spark of an idea, quietly flickering to life in the stillness of the mind. Like the dynamo effect in physics—where motion creates magnetic fields—this spark doesn't stay contained. With momentum, it grows, generating waves of influence and transformation. Thought, like the motion of charged particles, becomes a force that shapes the world.

The Ripple Effect of Ideas

An idea, once shared, acts like a pebble dropped into a still pond. The ripples spread, touching shores far beyond their origin. But this isn't just abstract motion—these ripples shape systems, communities, and civilizations.

For me, the *Transcendent Blueprint Theory* began as one of those sparks—a simple, intuitive thought. At first, I shared fragments of it, unrefined and incomplete, with others. But instead of fostering discussion, the ideas were met with scrutiny and skepticism. That resistance was a shock, and it forced me to withdraw and reflect.

What I didn't realize at the time was that this friction wasn't the end of the idea—it was the beginning. Much like the dynamo effect, that resistance refined the motion of my thoughts, sharpening and strengthening the concept until it became something far greater than its initial form. What started as a vague notion evolved, through friction and persistence, into the cohesive framework of the *Transcendent Blueprint Theory*.

Friction as a Catalyst for Growth

No idea—or movement—gains strength in a vacuum. The friction I faced in the early days of exploring TBT parallels the challenges others encounter when they venture into new territory. Whether it's immigrants adapting to new lands or scientists refining theories under scrutiny, resistance is the force that transforms raw potential into enduring progress.

- *Example:* A scientist presenting a bold new theory may face criticism, but that same critique refines the idea, ensuring it withstands scrutiny.
- *Example:* The immigrants who face societal resistance in their new home adapt and thrive, ultimately reshaping industries and communities for the better.

For me, the friction of early scrutiny opened a door to deeper understanding. It forced me to look harder, think broader, and articulate more clearly. The result wasn't just a refined idea—it was a revelation of the universal principles that connect all systems.

Thought as a System of Motion

Like physical motion, ideas don't merely ripple—they collide with existing systems, creating friction that sparks transformation. Whether it's the magnetism of a leader's bold actions or the quiet persistence of someone working through resistance, motion creates change, and resistance strengthens it.

For the Transcendent Blueprint Theory, the friction I encountered wasn't a barrier—it was the catalyst that brought clarity to the concept and shaped it into the framework it is today. That same dynamo effect of thought applies to everyone. Resistance refines. Friction sharpens. Motion, no matter how small, creates change.

The dynamo effect isn't just theoretical—it's personal. For me, it transformed an idea into a life's work. For others, it's the engine of progress, shaping lives, communities, and even nations. The friction you encounter isn't a failure—it's the push that refines motion into momentum, and motion into transformation.

3. Beyond the Tangible: Physics and Life's Intangibles

Physics often focuses on the measurable—the motion of planets, the properties of matter, or the transformation of energy. But its principles transcend the physical, shaping the intangible aspects of human life: emotions, relationships, and the evolution of societies. These forces, though unseen, are no less real and follow the same universal principles outlined in the *Transcendent Blueprint Theory* (TBT).

Gravity: The Invisible Threads That Bind Us
Gravity is more than a force—it is a principle of connection, pulling objects toward each other and holding systems in balance. In the same way, relationships act as gravitational forces in our lives, creating orbits of stability and influence.

> - *Example:* A mentor's influence parallels the pull of gravity. They provide a steadying force, grounding their mentee while encouraging them to reach higher, much like the gravity of a planet holding a moon in orbit. Without that connection, chaos or drift might ensue.

In families, workplaces, and communities, these gravitational forces manifest as bonds of trust, loyalty, and shared purpose. They are the invisible threads that keep individuals connected to larger systems, preventing isolation and fostering collective growth.

Entropy: The Catalyst for Renewal
The Second Law of Thermodynamics tells us that systems tend toward disorder, or entropy. While this principle often conjures images of decay, it is also a powerful catalyst for renewal and creativity. When the old systems begin to unravel, the space for new ones emerges.

> - *Example:* In the chaos of an economic downturn, innovation often thrives. Individuals and systems, pushed by necessity, adapt and create solutions that would never have been envisioned in times of stability. The disorder becomes the fertile ground for transformation.

This principle mirrors human resilience. Life's challenges—whether personal, professional, or societal—disrupt the status quo. But from the ashes of disorder, new opportunities arise, fueled by the same universal forces that govern energy and matter.

Motion and Magnetism: The Dynamics of Influence
The interplay of motion and magnetism in physics provides another lens to understand the human experience. Just as moving charged particles generate magnetic fields, our actions and decisions create ripples of influence, shaping the world around us.

- *Example:* A passionate leader, like a charged particle in motion, draws others into their magnetic field. This pull inspires collective action, while resistance sharpens their ideas and drives refinement. Similarly, personal motion—such as pursuing goals or embracing change—creates momentum that attracts opportunities.

Physics as a Bridge Between Realms

Physics doesn't just describe the tangible—it offers a blueprint for understanding the intangible forces that shape human existence. From the gravitational pull of relationships to the creative potential of chaos, the principles of motion, entropy, and connection illuminate the invisible threads that bind us to one another and to the systems we inhabit.

This bridge between the measurable and the immeasurable reveals the universality of the *Transcendent Blueprint Theory*. It shows us that the forces shaping galaxies are the same forces shaping human lives—proof that the cosmos and the soul are more connected than we ever imagined.

4. Parallels Across Disciplines: Physics as a Universal Language

Physics gives us a framework to understand the tangible world—forces that move mountains, bind planets, and fuel stars. But its reach extends far beyond atoms and energy. Through the lens of the *Transcendent Blueprint Theory* (TBT), we see how these principles—**cycles**, **motion**, **resistance**, **transformation**, **connection**, and **reciprocity**—echo across biology, thought, and society, creating a cohesive narrative of existence.

Cycles: Rhythms That Sustain Balance

Cycles are the foundation of all systems in physics. Energy, matter, and forces move through predictable rhythms, ensuring stability, balance, and renewal.

- **In Nature**: Planets orbit stars, tides rise and fall, and energy oscillates through waves. These cycles create consistency and sustain progress.

- *Example:* The Earth's orbit around the Sun not only creates seasons but ensures the balance of energy and life on the planet.

- **In Energy**: Thermodynamic cycles show how energy transforms but never disappears, powering systems like engines and ecosystems.
- *Example:* In a steam engine, energy cycles through phases—heat transforms water into steam, steam moves pistons, and energy dissipates as heat, ready to repeat.
- *TBT Insight:* Cycles in physics mirror nature's rhythms and human systems. Whether in planetary motion or economic booms and recessions, cycles sustain balance and prepare systems for renewal.

Motion: Energy That Sustains Progress

Motion is life's heartbeat, an unrelenting push forward. In the physical world, motion governs everything, from planetary orbits to the rush of a river carving canyons.

- **In Ecosystems**: Motion drives survival and balance. A predator chasing prey sharpens its skills, while the prey grows stronger and faster, ensuring balance.
- *Example:* When wolves were reintroduced to Yellowstone, their motion triggered a cascade. Elk fled overgrazed areas, trees returned, and even rivers shifted course as nature recalibrated its balance.
- **In Human Lives**: A person in motion—a parent rushing through their kitchen, a leader mobilizing a team—creates ripples of growth and change.
- *TBT Insight:* Motion ensures that systems adapt, progress, and thrive, much like celestial forces shape the cosmos.

Resistance: The Force That Shapes Us

Resistance stabilizes motion and transforms energy into refined strength. Friction and opposition act as sculptors, ensuring growth is purposeful.

- **In Nature**: A cactus thrives under resistance—scorching sun and arid soil. Its adversity forces it to store water, grow spines, and become a beacon of life.

- *Example:* Trees on windswept cliffs grow their roots and strengthen their wood, using resistance to survive and grow stronger.
- **In Human Lives**: Resistance refines systems, sharpening ideas and strengthening resolve.
- *Example:* Immigrants, pushed by hardship, transformed adversity into opportunity, building communities, industries, and nations.
- *TBT Insight:* Resistance is not the enemy of motion; it is the force that ensures progress aligns with strength, resilience, and purpose.

Transformation: Change That Defines Survival

Transformation is the soul of progress, turning potential into reality and chaos into order. In physics, energy changes form but is never lost.

- **In Nature**: Energy transforms through ecosystems, such as sunlight becoming food through photosynthesis.
- *Example:* A mangrove tree stabilizes coasts, shelters marine life, and transforms its surroundings into thriving ecosystems.
- **In Society**: Transformation emerges from challenges. Scarcity sparks innovation, and adversity births strength.
- *Example:* During the Great Depression, communities turned despair into progress, innovating through hardship.
- *TBT Insight:* Transformation connects survival with renewal, showing how systems evolve when they align with the universal flow of change.

Connection: The Web That Binds Us All

Connection is the invisible force that weaves systems into cohesive wholes, from planetary orbits to human relationships.

- **In Physics**: Gravity connects objects, holding planets in orbit and shaping galaxies.
- *Example:* The gravitational dance between Earth and Moon creates tides, which sustain ecosystems and life.

- **In Human Lives**: Emotional bonds, like gravity, create stability while enabling growth.
- *Example:* A parent's love grounds a child while launching them into independence, much like gravity steadies an orbit.
- *TBT Insight:* Connection is the thread that integrates systems, fostering resilience and harmony across the physical, biological, and human realms.

The Bridge Forward: Physics as a Universal Language

Physics reveals the foundational principles—**cycles**, **motion and resistance**, **transformation**, **connection**, and **reciprocity**—that govern existence. These forces are not confined to the physical world but extend into biology, ecosystems, thought, and society.

As we move forward, we will explore how these principles continue to shape the systems of **thought and human progress**, where ideas and relationships mirror the universal forces that sustain the cosmos.

A Universal Blueprint: Bridging the Seen and Unseen

Physics, at its heart, is the study of the forces that shape reality. But these forces aren't limited to the tangible; they ripple into every corner of existence, touching biology, thought, and society. The predator's pursuit mirrors the entrepreneur's hustle. The cactus's resilience reflects the immigrant's journey. The gravitational bond between planets finds its parallel in the love between parent and child.

Through the *Transcendent Blueprint Theory*, we see that motion, resistance, transformation, and connection are not isolated forces. They are the language of life, a universal story told through particles and people, ecosystems and ideas. By understanding physics as more than equations and matter, we uncover the threads that unite us with the cosmos—and with one another.

Conclusion: Physics as the Blueprint for Connection

Physics is not confined to equations or the motion of particles; it is the silent architect of existence, revealing the forces that bind us to one another and to the cosmos. Through motion and resistance, transformation, cycles, reciprocity and connection, it provides a universal language—one that transcends disciplines and bridges the tangible with the intangible.

The principles we see in physics ripple through every facet of life. They shape the orbits of planets, the growth of ecosystems, the evolution of thought, and the dynamics of human relationships. These forces, though often invisible, are the scaffolding of existence—a reminder that the smallest motion can spark transformation and the most subtle connection can stabilize entire systems.

As we move forward, we will shift our focus to thought and society, delving into the invisible frameworks that govern human interaction and innovation. Physics has shown us the mechanics of existence; now we turn to the dynamo of the human mind and the collective currents of society. Together, they reveal the blueprint of connection, growth, and transformation—a design that unites all systems into a cohesive whole.

Chapter 7: Thought — The Blueprint of the Mind

1. Introduction: Thought as the Dynamic Blueprint

Thought exists at the intersection of the tangible and intangible, a force that shapes both individual lives and collective realities. Invisible yet pervasive, it behaves like a living system, following the same universal principles that govern ecosystems, galaxies, and economies.

Through the lens of the *Transcendent Blueprint Theory* **(TBT)**, thought reveals its adherence to the principles of cycles, reciprocity, transformation, motion and connection. These universal laws allow thought to change, connect, and manifest its potential in the material world.

Thought is not chaotic or random; it is the seed of action, the architect of reality, and the dynamic blueprint that bridges the abstract with the tangible.

2. Thought as Energy: The Invisible Force

Thought is a form of energy, unseen but influential, operating much like the physical forces of the universe.

- **Dynamic Flow**: Like electricity flowing through a circuit, thoughts gain momentum through focus, repetition, and intent.
  - *Example:* A single, focused idea can lead to innovations or movements that ripple through society.
- **Transformation**: Thought transforms the abstract into the tangible, turning ideas into words, actions, and legacies.
  - *Example:* A vision for a better future becomes a technological breakthrough or a social movement.

Thought mirrors the transformation and flow of energy in physical systems, illustrating how the intangible gives rise to the tangible.

3. Universal Principles Applied to Thought

The principles of the *Transcendent Blueprint Theory* (TBT)—**cycles, reciprocity, transformation, motion and resistance,** and **connection**— govern thought as much as they do physical and biological systems. Thought is dynamic, moving, evolving, and influencing the world around it.

Cycles: The Rhythm of Thought

Thoughts move in cycles, from their creation to reinforcement, decline, and eventual renewal. These rhythms shape behavior, beliefs, and progress.

- *Example:* A recurring idea evolves into a belief system, influencing identity and action over time. When reframed or challenged, that idea enters a new cycle of growth.
- **Parallels in Nature**: Like water cycling through evaporation, rain, and flow, thoughts sustain momentum by moving through phases of creation, repetition, and transformation.

Reciprocity: The Exchange of Ideas

Reciprocity is the lifeblood of thought. Ideas grow when shared, challenged, and refined through interaction.

- *Example:* A solitary idea becomes richer through dialogue, debate, and collaboration, much like symbiotic relationships in ecosystems.
- *TBT Insight:* Thought thrives on reciprocal exchange. Just as ecosystems rely on mutual contributions, the sharing and refinement of ideas sustain intellectual and societal growth.

Transformation: Thought in Motion

Thought transforms as it interacts with the world. Ideas evolve, doubts become clarity, and intangible visions turn into tangible realities.

- *Example:* A fleeting idea transforms into a plan, a goal, or a movement. Doubt gives way to confidence, much like energy shifting between potential and kinetic states.
- *TBT Insight:* Transformation ensures that thought remains dynamic and adaptable, aligning with the universal law that change sustains progress.

Motion and Resistance: The Push and Pull of Thought

Thought is not passive; it moves, influences, and encounters resistance. Like physical forces, thoughts attract, repel, and refine the systems around them.

- **Motion**: Focused thought gains momentum, spreading through actions, relationships, and environments.

- *Example:* A leader's clear vision draws allies, creating momentum that propels teams and ideas forward.
- **Resistance**: Opposition challenges and strengthens thought, testing its foundation and refining its form.
- *Example:* A controversial idea may face resistance, but this friction sharpens its clarity and impact, ensuring it endures.
- *TBT Insight:* Motion drives thought into action, while resistance ensures its refinement and strength. Together, they shape thought into a powerful force of influence and growth.

Connection: Thought as the Binding Thread

Connection links thoughts to actions, individuals to ideas, and society to progress. It is the invisible thread that allows thought to bridge the abstract with the tangible.

- *Example:* A shared idea connects communities, movements, and generations, creating alignment toward a shared purpose.
- **Parallels in Nature**: Like mycorrhizal networks connecting plants underground, thoughts form networks of influence, spreading and sustaining growth.
- *TBT Insight:* Connection integrates thought across time and space, allowing ideas to resonate, align, and bring about transformation.

Thought: The Dynamic Blueprint of Reality

Thought adheres to the same universal principles—**cycles**, **reciprocity**, **transformation**, **motion and resistance**, and **connection**—as physical and biological systems. It is a living force, flowing, evolving, and shaping the world with every interaction.

4. Thought as a Living System

Thought behaves like a living system, reflecting the universal principles of growth, division, adaptation, and resilience. It mirrors the natural processes found in ecosystems, biology, and even physics, evolving as it interacts with its environment.

Growth and Division
Thought grows when nurtured and divides into new insights, much like cells in biology or branches on a tree. A single thought can expand, adapt, and inspire networks of interconnected ideas.

- *Example:* A scientist's hypothesis begins as a single idea but branches into new discoveries as research progresses. Each "branch" strengthens the system, much like how mycelial networks expand underground, spreading nutrients to sustain the forest.
- *TBT Insight:* Thought thrives on cycles of growth and division. Like natural systems, it diversifies to sustain innovation and resilience.

Adaptation and Resilience
Thought adapts when challenged, transforming limitations into opportunities. Just as ecosystems respond to environmental pressures, thought evolves to overcome adversity.

- *Example:* During societal upheavals, innovative solutions often emerge—such as wartime technological breakthroughs or economic recoveries born from hardship.
- **Parallels in Nature**: A tree exposed to strong winds grows reaction wood for strength. Similarly, resilient thoughts refine themselves under pressure, becoming clearer, stronger, and more impactful.
- *TBT Insight:* Adaptation and resilience are the natural responses to resistance. Thought behaves as a living system, growing stronger and more dynamic when challenged.

5. Parallels Across Disciplines
Thought, much like energy or ecosystems, operates within universal principles. Its behavior connects disciplines, demonstrating how these laws transcend boundaries and shape all systems.

Biology: Thought as a Cognitive Ecosystem
Thought mirrors ecosystems—ideas compete for resources like organisms, shaping the mental "landscape." Productive thoughts promote growth, while harmful patterns can overtake and disrupt balance.

- *Example:* Positive, productive thoughts dominate when nurtured, creating clarity and purpose. Conversely, negative thought patterns—akin to invasive species—can overrun the mind, consuming attention and energy.
- *TBT Insight:* Like ecosystems, the health of thought relies on diversity, balance, and reciprocal exchanges. Mental resilience stems from cycles of renewal, pruning, and adaptation.

Physics: The Magnetism of Thought

Thought operates like magnetic fields, creating influence that attracts or repels people, actions, and events. The energy of a thought radiates outward, shaping its environment.

- *Example:* A charismatic idea can inspire and unify communities, pulling people together like gravitational forces. Conversely, divisive or fear-driven ideas create resistance, pushing others away and destabilizing systems.
- *TBT Insight:* Thought, like magnetism, creates motion and resistance. Its energy reshapes reality, connecting systems and influencing outcomes through attraction or opposition.

Economics: The Currency of Thought

Thought acts as capital, exchanged and transformed to generate progress. Ideas drive innovation, investments, and systems of value.

- *Example:* A startup begins as a simple idea. With shared vision, investment, and collaboration, that thought scales into an industry, reshaping economies and societies.
- *TBT Insight:* Thought follows cycles of investment, transformation, and return. Ideas, like currency, grow when exchanged, refined, and applied to create greater value.

6. Reciprocity and Contagion in Thought

Thought spreads like a living force, operating through **reciprocity** and **contagion**. It flows through networks, creating ripples of influence that expand exponentially.

Ripple Effects: The Power of Reciprocity

Thought grows when given, shared, and received. Like nature's cycles of reciprocity—where trees share nutrients through roots—thought thrives when it engages with others.

- *Example:* A revolutionary idea shared in a small group sparks innovation, spreading outward and creating movements that reshape entire industries or cultures.
- *TBT Insight:* Reciprocity ensures thought remains dynamic and resilient. By exchanging and refining ideas, systems grow stronger, much like mutual exchanges sustain ecosystems.

Contagion: Thought as a Living Current

Thought behaves like a contagion—spreading rapidly, influencing others, and reshaping systems. Ideas, like emotions, are infectious, sparking momentum and collective action.

- *Example:* A viral idea on social media can ignite global movements, from social justice campaigns to technological revolutions.
- **Parallels in Nature**: Just as fire spreads through a forest, igniting new growth, thought can "catch" in networks, driving transformation and renewal.
- *TBT Insight:* Contagion amplifies the power of thought, accelerating cycles of transformation and connection across systems.

7. Thought in Cycles: Creation, Reinforcement, Renewal

Thought follows a cyclical journey, ensuring its adaptability and growth. These cycles mirror the natural rhythms found in ecosystems and physics, aligning thought with universal patterns of renewal.

Creation

Thought begins with curiosity, inspiration, or necessity. It emerges as a spark—an idea ready to take shape.

- *Example:* A challenge inspires innovation, like a spark igniting a flame.

Reinforcement
Repetition strengthens thought, embedding it as a belief, habit, or cultural norm. Just as cycles in nature sustain systems, reinforcement preserves thought's influence.

> • *Example:* Daily repetition of empowering thoughts transforms self-perception and behavior, much like steady rainfall nourishes growth.

Renewal
Old thoughts are revisited, reshaped, or replaced to align with new realities. Renewal ensures thought remains dynamic, adaptable, and in harmony with its environment.

> • *Example:* A limiting belief changes through introspection and reframing, empowering new growth. This mirrors forests renewing through fire, clearing space for new life to emerge.
>
> • *TBT Insight:* Thought, like nature, relies on cycles of creation, reinforcement, and renewal. These rhythms ensure its survival, adaptability, and continued progress.

Thought behaves as a living system, following the universal principles of **cycles, reciprocity, transformation, motion and resistance**, and **connection**. It spreads, evolves, and influences, shaping the tangible world with its unseen energy. Understanding thought through the lens of the TBT allows us to see it as a dynamic, powerful force—one that bridges the abstract and material, the individual and collective, the past and future.

8. Conclusion: Thought as the Invisible Blueprint

Thought is not merely a fleeting process; it is the dynamic force that drives progress, innovation, and connection. Governed by universal principles, thought grows, transforms, and influences, acting as the invisible architect of reality.

Through the **Transcendent Blueprint Theory**, we see thought as a system in alignment with physical and biological laws. It bridges the intangible and the material, proving that even the unseen adheres to the universal laws that govern existence.

As we transition to the next chapter, we will explore how societies embody these principles, revealing the collective blueprint of human connection and progress.

Chapter 8: Humanity Introduction to the Universal Blueprint

1. Introduction: Humanity's Place in the Blueprint

Humanity, like all systems, exists within the universal blueprint. We are governed by the same principles of cycles, reciprocity, transformation, motion and connection that sustain nature and the cosmos. Yet, our capacity for reflection and deliberate action sets us apart, creating the perception that we are somehow outside or above these laws.

This chapter explores humanity through the lens of the TBT, showing how universal laws manifest in our thoughts, behaviors, and societies. By understanding ourselves as participants in the blueprint, we can move from unconscious expression to intentional alignment, deepening our role within the web of life.

2. Cycles in Humanity: Patterns of Renewal and Resistance

Humanity operates within cycles—of growth, learning, renewal, and decline. These patterns, far from disruptions, are natural expressions of the blueprint, reflecting the rhythms that sustain all systems.

- **The Cycles of Human Development**:
- At an individual level, humans grow through stages of learning, reflection, and transformation. These cycles mirror nature's rhythms, where dormancy or rest precedes growth.
- *Example:* Personal growth often emerges from challenges, reflecting the cyclical relationship between adversity and renewal.
- **Societal Cycles**:
- Societies rise, evolve, and sometimes collapse, following larger patterns of creation, resistance, and transformation.
- *Example:* Periods of social unrest often precede significant cultural or political advancements, illustrating that breakdowns pave the way for breakthroughs.
- *TBT Insight:* Humanity's cycles are not random but governed by the universal principle of renewal. Connection integrates

these cycles, tying individual growth to collective progress and ensuring alignment with the larger rhythms of existence.

3. Reciprocity in Humanity: Balancing the Give and Take

The universal law of reciprocity governs all human systems, ensuring balance through exchange—whether between individuals, communities, or humanity and its environment. While disruptions may occur, the blueprint ensures all systems self-correct over time.

- **Interpersonal Reciprocity:**
- Relationships thrive on balanced exchanges of energy, support, and understanding. Imbalances—such as neglect or exploitation—create tension, prompting feedback loops that restore balance.
- *Example:* A friendship where one person takes without giving collapses, forcing renewal or reconnection.
- **Global Reciprocity:**
- Humanity's actions, such as resource extraction without replenishment, trigger feedback mechanisms that demand adaptation and restoration.
- *Example:* Overuse of farmland depletes soil, but practices like crop rotation or regenerative farming restore balance and ensure sustainability.
- *TBT Insight:* Reciprocity highlights the interconnected flow of giving and receiving. These exchanges create stability, allowing humanity to align with universal patterns of balance and renewal.

4. Transformation: Humanity's Creative Force

Humanity's ability to transform raw materials, experiences, and ideas into something new reflects the universal principle of transformation. This creative force allows systems to adapt, evolve, and align with the blueprint.

- **Personal Transformation:**
- Individuals turn challenges into growth, converting adversity into strength, purpose, and connection.
- *Example:* Overcoming personal hardship builds resilience, leading to insights that ripple outward to benefit others.

- **Societal Transformation:**
- Collectively, humanity channels creativity into changing systems, reshaping structures that no longer serve their purpose. Even collapse fosters conditions for new growth.
- *Example:* The Industrial Revolution, while disruptive, transformed economies and societal progress. Today, technological revolutions may carry us to a similar threshold of change.
- *TBT Insight:* Transformation connects destruction to renewal. Humanity's creativity ensures that even challenges or collapse become catalysts for growth, reflecting the blueprint's ability to reshape systems for alignment and progress.

5. Motion and Magnetism: The Forces That Drive Humanity

Motion and magnetism govern human systems by driving progress and fostering attraction or resistance. These forces keep humanity dynamic, responsive, and in constant alignment with the universal blueprint.

- **Magnetism in Human Relationships:**
- Thoughts, emotions, and actions create energy that pulls people together or pushes them apart. This magnetic force shapes communities, cultures, and collective action.
- *Example:* A shared vision for equality unites individuals across barriers, creating a magnetic pull toward a common goal. Fear or division, by contrast, repels and disrupts connection.
- **Motion in Societal Progress:**
- Societies move through cycles of momentum and resistance. Friction, though challenging, sparks creativity and drives transformation.
- *Example:* The abolitionist movement generated the resistance needed, creating forward momentum for societal progress.
- *TBT Insight:* Motion propels systems forward, while magnetism fosters alignment. Together, they ensure human systems remain dynamic, purposeful, and resilient in their pursuit of balance and growth.

6. Connection: Humanity's Unifying Thread

Connection is the invisible thread that binds individuals, systems, and ideas into a cohesive whole. It fosters collaboration, aligns shared purpose, and ensures progress through networks of understanding and unity.

- **Interpersonal Connection**:
- Connection strengthens relationships, enabling collaboration and shared growth.
- *Example:* A mentor's guidance connects experience to potential, creating a ripple effect that empowers future generations.
- **Cultural and Societal Connection**:
- Shared values and goals connect individuals into communities. These bonds align societies, allowing progress to emerge from collective effort.
- *Example:* Many movements, rooted in shared principles, unify diverse groups to challenge the imbalance creating force to lead systems back to alignment.
- **Connection Across Systems**:
- Humanity builds networks of connection through tools, ideas, and shared knowledge, fostering progress that mirrors the balance seen in natural systems.
- *Example:* The printing press connected ideas across borders, creating intellectual revolutions that reshaped societies and elevated human potential.
- *TBT Insight:* Connection transcends boundaries, weaving isolated elements into thriving systems. It ensures alignment with the universal principles of growth, balance, and renewal, allowing humanity to evolve collectively.

Humanity, seen through the lens of the *Transcendent Blueprint Theory*, aligns seamlessly with the universal principles—**cycles, reciprocity, transformation, motion and resistance, and connection**. These forces govern personal growth, societal evolution, and humanity's interactions with its environment.

Far from random or chaotic, humanity's challenges, adaptations, and progress reflect a greater order—a living system bound by laws that transcend disciplines and align all systems, physical or intangible, into patterns of balance and renewal.

6. Humanity as a Biological System

Through the lens of the TBT, humanity reflects the same principles that govern biological systems. Our behaviors and societal patterns mimic the dynamics of growth, competition, and adaptation.

- **Source-to-Sink Dynamics**:
    - Just as ecosystems balance sources (energy creators) and sinks (energy consumers), humanity mirrors this pattern in its distribution of resources and power. Imbalances, such as the concentration of wealth or the suppression of populations, create instability that eventually self-corrects.
    - *Example:* Authoritarian regimes often create "sinks" by suppressing dissent, but these imbalances generate the conditions for eventual resistance and renewal.

Humanity, as a biological system, operates within the same cycles of adaptation, growth, and renewal that define all life.

7. Humanity's Alignment with the Blueprint

Humanity's alignment with the universal blueprint is not a choice but a reality. Every action, thought, and system we create reflects the principles of the TBT, even when these manifestations appear disruptive or destructive.

- **Natural Alignment**:
    - Humanity cannot exist outside the blueprint. What we perceive as "misalignment" is simply the blueprint expressing itself through imbalance, correction, and adaptation.

- **Conscious Participation**:
    - While our alignment is inevitable, our awareness of it determines whether we engage with intention or resistance. By understanding the principles of the TBT, we can consciously participate in cycles, reciprocity, and transformation, amplifying their positive impacts.

Humanity is inherently aligned with the blueprint; our challenge is not alignment itself but the conscious application of these principles in our actions and systems or acknowledgement of them when they occur.

8. Conclusion: Humanity as Co-Creator

Humanity's unique capacity for reflection and creativity makes us dynamic within the universal blueprint. We do not operate outside its principles, but our awareness of these laws allows us to shape their manifestations with intention and purpose.

By understanding humanity through the lens of the *Transcendent Blueprint Theory,* we see ourselves not as disruptors but as participants in the ongoing cycles of growth, renewal, and connection. Our role is not to force alignment but to embrace it, creating systems and structures that are adaptable to its rhythms. Advancements that benefit the whole require not only a diversity of thought but also the intentional use of our words—the seeds that shape reality and direct the flow of progress.

Chapter 9: The Gravity of Words

Introduction: Words as Universal Forces
Gravity stabilizes the cosmos, binding planets, stars, and galaxies in harmonious motion. But gravity's influence is not confined to the physical realm—it extends into the intangible. Words, like physical forces, possess a kind of gravity. They carry weight, create attraction or repulsion, and ripple outward, influencing systems far beyond their origin.

Through the lens of the *Transcendent Blueprint Theory* (TBT), words emerge as dynamic participants in the universal laws of cycles, reciprocity, transformation, motion, and balance. They are not merely tools of communication but forces of alignment and creation, capable of shaping the intangible realms of thought and emotion as well as the tangible realities of relationships, societies, and structures.

This chapter explores the gravitational power of words, revealing their role in stabilizing systems, catalyzing change, and perpetuating influence. By understanding words through the TBT, we uncover their profound ability to create, connect, and transform—aligning the abstract with the universal blueprint that governs all existence.

2. The Tangible Weight of Words
Words, while intangible, exist within physical and energetic realities that reflect the universal laws of the *Transcendent Blueprint Theory*. Like celestial bodies or atomic particles, words carry energy, exert influence, and create ripples that align with the principles of motion, transformation, and contagion. No one understood this more than Jesus Chris himself.

A. The Physics of Words
Words are energy in motion, expressed through sound waves that interact with the physical environment.

- **Energy in Motion**:
  - Spoken words generate sound waves, carrying energy through vibrations. These waves displace particles, creating measurable frequency and amplitude that interact with listeners and surroundings.
  - *Example:* A low, calm voice resonates as soothing, while a sharp, loud tone evokes urgency or alarm.

- **Interaction with the Physical World**:
    - Just as gravity acts on physical objects, sound waves produced by words are subject to gravitational forces, though this effect is negligible in perception.
    - Research on sound vibrations influencing materials like water and brainwaves suggests that words physically impact their environment, reflecting their energetic presence.

Words are more than abstract concepts—they are physical phenomena that move and transform energy, aligning with the universal law of motion.

B. Words as Mass and Gravitational Pull

Through the TBT, we see words as entities with metaphorical "mass," reflecting their energy, significance, and ability to draw people into alignment.

- **Mass and Influence**:
    - Words with deeper meaning or emotional charge act as centers of gravity, pulling individuals and communities toward shared understanding or action.
    - *Example:* Phrases like "freedom," "justice," or "hope" have immense gravitational pull, shaping social movements and cultural identities.

- **Amplification Through Repetition**:
    - Repeated words or phrases accumulate metaphorical "mass," amplifying their influence much like adding weight increases gravitational force.
    - *Example:* Mantras, slogans, and repeated affirmations solidify ideas and create enduring systems of belief or action.

- **Binding Energy in Words**:
    - Words operate like the atomic binding energy that holds particles together, creating stability and structure in human interactions and ideas.
    - *Example:* A heartfelt apology or commitment stabilizes relationships, much like gravity stabilizes planetary orbits.

Words gain mass and energy through repetition and emotional resonance, reflecting the universal law of transformation as they shape thoughts, actions, and systems.

C. The Universal Blueprint in Words
Words embody the universal principles of the TBT in the following ways:

1. **Motion: Words, as sound waves, carry energy through systems, creating ripples of influence.**
2. **Transformation: Like atomic energy, words transform abstract thoughts into tangible effects, driving actions and change.**
3. **Contagion: Words spread like chain reactions, influencing individuals and systems far beyond their initial expression.**
4. **Balance: Words stabilize relationships and communities, creating harmony when used intentionally.**
5. **Cycles**: Words often return to their source, creating feedback loops that sustain positivity or perpetuate negativity.

The energy and mass of words align them with the universal laws, amplifying their power to create, connect, and transform.

3. Words as Stabilizers and Connectors
Through the lens of the *Transcendent Blueprint Theory,* words act as forces of alignment, creating stability and connection within systems. Just as gravity binds celestial bodies into harmonious orbits, words anchor relationships, communities, and ideas, fostering equilibrium and shared purpose.

A. Words as Bridges
Words function as bridges between individuals, enabling understanding, empathy, and alignment across differences.

- **Facilitating Connection**:
  - Words build pathways of communication, fostering bonds that enable collaboration and shared goals.
  - *Example:* A heartfelt apology or sincere expression of gratitude can heal rifts, creating emotional alignment and trust.
- **Building Understanding**:

- Words clarify intentions and provide the structure for mutual respect and cooperation.
- *Example:* A diplomat's precise language can defuse tension, paving the way for peace.

Words as bridges align with the TBT principle of connection, linking systems and fostering cohesion across boundaries.

B. Stabilizing Systems

Words are stabilizers within relationships, communities, and organizations, functioning like gravitational forces that maintain balance and prevent fragmentation.

- **Reinforcing Trust Through Truthful Communication:**
  - Clear, consistent, and truthful language acts as an anchor, providing stability in times of uncertainty or change.
  - *Example:* Leaders who communicate transparently inspire confidence, much like gravity holding systems together during motion.

- **The Friction of Miscommunication:**
  - Ambiguous or thoughtless language destabilizes systems, introducing friction that disrupts alignment and balance.
  - *Example:* A vague or poorly chosen phrase in a high-stakes negotiation can unravel trust, leading to disconnection or conflict.

- **Words as Gravitational Centers:**
  - Certain words and phrases serve as stabilizing anchors, rallying individuals or communities around shared values.
  - *Example:* National mottos like "E pluribus unum" unify people under a common identity, much like a gravitational center stabilizes an orbiting system.

Words embody the TBT principle of balance, functioning as stabilizers that maintain harmony within dynamic systems when used intentionally and truthfully.

Words as Stabilizers and Connectors

Words, when aligned with the universal principles of the *Transcendent Blueprint Theory,* stabilize and connect systems. They act as bridges, creating understanding and cohesion, and as anchors, reinforcing trust and alignment. Their power to stabilize relationships and communities reflects the universal law of balance, ensuring that systems thrive when supported by thoughtful and intentional communication.

4. The Ripple Effect of Gravitational Words

Words, like gravitational forces, extend their influence far beyond their immediate context. They ripple outward, shaping thoughts, behaviors, and systems over time and space. Through the lens of the TBT, these ripples embody the universal laws of cycles, contagion, and transformation, showing how words perpetuate energy and create enduring impacts.

A. Proximity of Influence

- Words spoken in close relationships exert the strongest gravitational pull, shaping emotions, thoughts, and actions with immediacy and intensity.
    - *Example:* A teacher's encouragement can ignite a student's lifelong passion, while careless criticism can create lasting emotional scars.

B. Historical Ripples

- Some words transcend their original context, acting as gravitational centers that anchor values, ideas, and movements for generations.
    - *Example:* The phrase "We the People" continues to ripple through history, embodying democratic ideals and uniting diverse communities under shared principles.

C. Expanding Influence

- As words spread outward, their effects may lessen in intensity but broaden in scope, touching new systems and creating indirect impacts.
    - *Example:* A powerful quote shared on social media can inspire action in distant parts of the world, illustrating the interconnectedness of human systems.

The ripple effect of words reflects the TBT's universal laws, particularly cycles and contagion, as their influence perpetuates and amplifies across interconnected systems over time.

D. Jesus: Words as Divine and Transformative Forces

- Among the most profound examples of words' power is found in the life and ministry of **Jesus Christ**. His use of words transcended human understanding, bridging the physical, spiritual, and eternal. While one may say His use of words effectively was due to being the Son of God they would err in that statement. Because He too was a man. Possibly what can be gleaned is that TRUTH is a direct connector to the power of words creating more alignment with TBT principles.

- **Words as Creation**:

Jesus spoke with authority, using words to heal, restore, and bring life. His words embodied motion and transformation, carrying divine intent and aligning with universal principles.

- **Example:** When Jesus said, *"Lazarus, come forth,"* His words transcended the natural laws of death, transforming stillness into life. His spoken word turned potential into reality.

- **Words as Seeds of Truth**: Jesus' parables revealed the cyclical nature of words. Spoken truth, like seeds, takes root in receptive hearts, growing into faith, understanding, and action.

  - **Example:** The Parable of the Sower emphasizes that words (seeds) must land on fertile ground to bear fruit, aligning with reciprocity and cycles. Those who "have ears to hear" are transformed by their growth.

- **Words as Connection**: Jesus' words were relational, connecting individuals to God, to one another, and to their higher purpose. His teachings fostered love, forgiveness, and unity, aligning human systems with divine principles.

  - **Example:** *"Love your neighbor as yourself"* embodies the law of reciprocity and connection, establishing balance and harmony in relationships.

- *TBT Insight:* Jesus' words reflect the universal blueprint. They demonstrate how words, when aligned with divine intention and universal laws, create, connect, and transform systems in ways that transcend time, space, and circumstance. I stop short on this as a whole book could be written of Jesus and His words. ie the mountain thrown into the see, He said "tell" or in other words "speak" to the mountain, or in another example the withered tree that He spoke to and cursed. We could go on to add the whole Bible if we wanted, full of examples of verbal blessings and curses.

5. Words and the Universal Blueprint

Words are active participants in the *Transcendent Blueprint Theory*, embodying its universal laws and transcending disciplines like physics, biology, sociology, and psychology. They function as dynamic forces, bridging the tangible and intangible, and grounding abstract ideas in physical and social realities.

A. Cycles

- Words create feedback loops, perpetuating energy through systems of communication and interaction.
  - *Example:* A simple compliment today can inspire positivity tomorrow, cycling energy back into the system.

B. Reciprocity

- Words reflect the energy of their intent, often returning to their source in the form of responses, actions, or relationships.
  - *Example:* Encouragement fosters trust and connection, while harsh words create resistance and division.

C. Transformation

- Words transform abstract thoughts into tangible actions and outcomes, reshaping systems and realities.
  - *Example:* A visionary speech transforms potential into collective action, inspiring structural and cultural change.

D. Contagion

- Words spread across systems, influencing individuals, communities, and global movements.

- *Example:* Viral slogans or shared ideas demonstrate the contagious nature of words, amplifying their impact across interconnected networks.

E. Balance

- Words stabilize relationships, systems, and communities by fostering harmony and alignment when used with care and intention.
  - *Example:* Diplomatic language defuses tension, restoring balance and creating pathways for collaboration.

Words embody the TBT's universal principles, serving as dynamic forces that shape, sustain, and transform systems. When aligned with these laws, words amplify their power to create, connect, and catalyze progress.

6. Speaking with Gravity

To speak with gravity is to recognize and intentionally harness the multifaceted power of words. Words are not passive; they carry energy, shape realities, and ripple through systems. Aligning them with the universal laws of the *Transcendent Blueprint Theory* amplifies their capacity to create, connect, stabilize, and transform.

A. Speak with Purpose

- Use words that add clarity, value, or connection. Every word spoken should be intentional, aligned with the energy you wish to create.
- Avoid language that divides, destabilizes, or diminishes, as such words disrupt cycles and create resistance.

Example: A leader who communicates transparently builds trust and alignment, while vague or divisive language fractures unity.

Purposeful words align with the TBT principle of motion, ensuring that energy flows constructively through systems.

B. Amplify Positive Ripples

- Speak to inspire, encourage, and uplift, creating cycles of trust, connection, and growth.
- Recognize that words ripple outward, influencing others far beyond their immediate context.

Example: A teacher's encouragement can inspire generations of students, demonstrating how positive ripples expand into larger systems.

Positive ripples align with the TBT principle of cycles, perpetuating energy that sustains and builds systems.

C. Anchor Trust and Truth

- Consistent, truthful language stabilizes relationships, much like gravity stabilizes celestial systems.
- Truthful words act as anchors, providing clarity and reliability in times of uncertainty or change.

Example: During a crisis, clear and honest communication fosters calm and focus, ensuring alignment and collective resilience.

Anchoring trust aligns with the TBT principle of balance, ensuring stability and harmony within systems.

D. Embrace Words as Builders and Catalysts

- Use words to build visions, reshape thought, and catalyze action. Recognize their potential to initiate tangible and intangible creation.
- Avoid destructive language that erodes hope, divides communities, or halts progress.

Example: A visionary statement like "Yes, we can" builds collective momentum, transforming individual energy into shared purpose.

Words as builders align with the TBT principle of transformation, turning potential energy into action and outcomes.

Speaking with Gravity

To speak with gravity is to align your words with the universal blueprint, ensuring they contribute to systems of connection, balance, and growth. Recognizing their capacity to build, ripple, and stabilize, you transform words into forces of creation, advancing the interconnected cycles of the TBT.

Conclusion: Words as Seeds of Connection

Words are not ephemeral; they are builders of reality. They align thought and action with the universal blueprint, embodying the principles of cycles, reciprocity, and transformation. Like gravity, words pull, stabilize, and connect, creating ripples that shape individuals, communities, and societies.

Every word carries the potential to grow into actions, relationships, and legacies that define the systems we inhabit. Just as gravity binds celestial bodies into harmonious orbits, words anchor humanity within the vast web of existence, fostering connection, balance, and progress.

But words are only one part of the story. Humanity itself is a system shaped by the energy of ideas, the gravity of words, and the actions they inspire. How we interact, collaborate, and align with universal principles determines not only the strength of our connections but also the systems we build and sustain.

In the next chapter, we will explore on a deeper level humanity's place within the universal blueprint, examining much closer how our shared existence reflects and challenges the laws that govern all systems. From relationships to societies, humanity's role is both to shape and be shaped by these universal truths, forging connections that ripple across generations.

Bonus "JESUS" Section

To lighten your mental load before revisiting Humanity again I offer you a detour of thought. The mention of Jesus in this chapter tells you quite where I stand. I believe He knew things that we did not and knew of this theory before anyone else. After all it is His design. But maybe you don't believe that, and that's ok. But I do, so enjoy seeing Jesus through the TBT lens. Here are the five core principles as it pertains to Jesus Christ and I often wonder as I write this book of the Blueprint, Theories and Laws, if He thought the same thoughts.

1. Motion: Forward Movement and Momentum

Jesus exemplified motion through His **purposeful actions**, catalyzing spiritual, social, and cultural transformation. His teachings and life were not passive but marked by **intentional movement** that inspired others to change and grow.

- **Physical Motion and Ministry**:

Jesus' ministry was characterized by movement—walking from town to town, teaching, healing, and spreading His message. His presence brought action wherever He went, creating ripples of transformation.

- *Example:* His journey to Galilee, Judea, and Jerusalem spread His teachings, gathering disciples and followers who continued His mission long after His crucifixion.

- **Momentum of Ideas**:

Jesus' words created a powerful **spiritual and social momentum** that transcended His time. His parables and teachings, like seeds, planted ideas that grew into global movements.

- *Example: "Love your neighbor as yourself"* sparked a shift in moral and social behavior that continues to influence cultures worldwide.

- **Push and Pull**:

Jesus' actions also created resistance—a natural consequence of motion. His challenges to religious and social norms generated opposition, but that resistance only sharpened the clarity and power of His mission.

- *Example:* His overturning of the tables in the temple confronted corruption, disrupting stagnation and pushing the system toward reform.

Jesus' motion was purposeful and aligned with truth, generating resistance that acted as a refining force. His forward momentum created ripples that spanned generations, demonstrating that aligned motion leads to enduring transformation.

2. Cycles: Renewal, Resurrection, and Perpetuity

Jesus' life and mission are grounded in cycles of **death, renewal, and rebirth**, aligning with the rhythms of the natural and spiritual world.

- **The Cycle of Teaching and Learning**:

Jesus repeatedly taught in **cycles**—through parables, lessons, and actions—reinforcing core principles until they were understood, internalized, and spread.

- *Example:* The Sermon on the Mount was not a one-time event; its truths continue to cycle through history, renewing generations with teachings of love, humility, and forgiveness.

- **The Crucifixion and Resurrection**:

The most profound expression of the **cycle of transformation** is found in Jesus' death and resurrection. His crucifixion symbolized the end of one phase, while His resurrection marked the renewal of life, hope, and faith.

- *Example:* Jesus' resurrection broke the cycle of sin and death, offering humanity a path to eternal renewal. This echoes nature's patterns of death and rebirth, like a seed that must die to produce new life.

- **Communion as a Cycle of Renewal**:

The practice of Communion reflects an ongoing cycle of remembrance and renewal, aligning believers with the sacrifice of Christ and the promise of redemption.

- *Example:* "*Do this in remembrance of me*" transforms a single moment into a recurring act that reinforces connection, community, and faith across time.

Jesus embodied the cycle of transformation—dying to bring new life, breaking stagnation to restore balance. His resurrection is not only a spiritual truth but a reflection of the universal principle that **renewal follows endings**, ensuring growth and progress.

Through motion, Jesus brought purposeful energy and transformation, generating momentum that reshaped the world. Through cycles, He demonstrated the universal rhythms of renewal and rebirth, offering humanity a path toward restoration. Together, these principles reflect the **dynamic balance** at the heart of the *Transcendent Blueprint Theory*, aligning Jesus' life and teachings with the same laws that govern all systems.

3. Reciprocity: The Balance of Giving and Receiving

Reciprocity is the universal flow of energy that ensures systems remain balanced, sustainable, and connected. In Jesus' life and ministry, reciprocity was not transactional but a dynamic exchange that fostered wholeness, unity, and renewal.

Giving Freely, Receiving Abundantly

Jesus exemplified reciprocity as an act of grace—giving without expectation, trusting that the cycles of love, provision, and healing would return in ways aligned with a higher purpose.

- **Miracle of Provision**: Jesus fed five thousand people with only five loaves and two fish. A small offering became an abundance that sustained thousands, illustrating that giving, no matter how limited it seems, fuels cycles of provision and trust.
  - *TBT Insight:* The universal law of reciprocity ensures that energy given with intention and alignment multiplies, sustaining both giver and receiver.
  - *Scripture: "Give, and it will be given to you. A good measure, pressed down, shaken together and running over, will be poured into your lap."* (Luke 6:38)
- **Teaching on Forgiveness**: Jesus tied the act of forgiving others to the ability to receive forgiveness, demonstrating reciprocity as a pathway to restoration.
  - *Example:* The Parable of the Unmerciful Servant highlights the imbalance that occurs when reciprocity is disrupted. A man forgiven of a great debt refused to forgive another, triggering the collapse of grace within the system.

Healing as Reciprocal Alignment

Many of Jesus' miracles involved both giving and receiving—a flow of faith, trust, and healing that brought people into alignment with the universal blueprint.

- *Example:* The woman who touched Jesus' cloak and was healed from years of suffering did so through faith. Jesus said, *"Your faith has made you well."* Her act of belief connected her to the energy of healing, demonstrating the reciprocity between intention (faith) and outcome (healing).
  - *TBT Insight:* Reciprocity in spiritual systems mirrors natural laws; the energy we put forth—whether through faith, kindness, or love—returns to us in ways that realign and renew.

Spiritual Reciprocity: Sowing and Reaping

Jesus often spoke of reciprocity as a law of sowing and reaping. What is planted—whether good or destructive—returns in kind, aligning with cycles of balance and consequence.

- *Scripture:* "A good tree cannot bear bad fruit, and a bad tree cannot bear good fruit." (Matthew 7:18)
- *Example:* In the Parable of the Sower, seeds scattered on fertile soil yielded abundance, while seeds on rocky or thorny ground produced little. This highlights that reciprocity thrives when systems are aligned, ready, and capable of receiving energy.

Jesus' life reflects reciprocity as the balanced flow of energy and intention. Giving leads to receiving, healing flows from faith, and restoration follows forgiveness. Systems flourish when reciprocity aligns with universal rhythms.

4. Transformation: Turning Potential into Renewal

Transformation is the act of turning raw potential into new realities—renewal, growth, and alignment with purpose. Jesus' ministry embodied transformation in its purest form, reshaping lives, systems, and perceptions to reflect universal truth.

Physical Transformation: Healing and Wholeness

Jesus' miracles often transformed the physical, restoring bodies to health and lives to wholeness. These acts were not merely displays of power but tangible manifestations of universal principles.

- *Example:* Jesus healed a paralyzed man, telling him to "take up your mat and walk." This physical transformation rippled outward, restoring the man's dignity, connection to society, and purpose.
  - *TBT Insight:* Transformation begins with potential. Systems realign and renew when energy is directed toward restoring balance and motion.
- **Raising the Dead**: Jesus' raising of Lazarus represents transformation at its most profound—life emerging from death, mirroring the universal law that renewal follows endings.
  - *Scripture:* "I am the resurrection and the life. The one who believes in me will live, even though they die." (John 11:25)

Spiritual Transformation: Changing Hearts and Minds

Jesus' teachings transformed internal systems—shifting beliefs, renewing minds, and restoring relationships. He invited people to see the world through the lens of love, forgiveness, and humility.

- *Example:* Zacchaeus, the tax collector, experienced radical transformation after encountering Jesus. Once greedy and exploitative, he pledged to return stolen wealth and give generously to the poor. His heart was reshaped, and his actions aligned with a new purpose.
  - *TBT Insight:* Transformation aligns systems with truth. Jesus' presence turned internal potential into external change, demonstrating the universal principle of renewal through alignment.

The Cross: Ultimate Transformation

The crucifixion and resurrection stand as the pinnacle of transformation—turning suffering into redemption, death into life, and despair into eternal hope.

- **The Shift**: The cross, a symbol of execution, was transformed into the ultimate representation of love, sacrifice, and renewal.
  - *TBT Insight:* Transformation is cyclical. Endings create conditions for new beginnings, and suffering—when aligned with purpose—becomes the catalyst for growth.

Jesus' life reveals transformation as a universal force that reshapes systems from within, aligning them with their highest potential. Physical, spiritual, and systemic change flows when energy is directed toward truth and renewal.

5. Connection: Uniting Systems Through Love and Purpose

Connection binds individuals, communities, and systems into cohesive, thriving networks. Jesus' ministry centered on connection—restoring relationships, uniting people with God, and building networks of love and purpose.

Connection to God

Jesus emphasized that humanity's ultimate connection is to God, the source of life and purpose. This connection sustains and aligns all systems.

- *Scripture:* "*I am the vine; you are the branches. If you remain in me and I in you, you will bear much fruit.*" (John 15:5)
- *TBT Insight:* Connection to the source fuels growth, resilience, and purpose. Systems thrive when they remain aligned with their foundational energy.

Restoring Broken Connections

Jesus actively sought out the marginalized, reconnecting them to society, dignity, and hope. His actions revealed connection as a force for healing and inclusion.

- *Example:* The Samaritan woman at the well was isolated by cultural and personal barriers. Jesus engaged her with compassion, restoring her sense of worth and connection to community.
- *TBT Insight:* Connection bridges divides, fostering balance and wholeness within systems.

Building Networks of Purpose

Jesus connected individuals into a collective—His disciples—entrusting them with a shared mission. This network carried His teachings forward, creating a legacy of transformation and alignment.

- **The Great Commission**: *"Go therefore and make disciples of all nations."* Jesus' command created a web of connection that transcended borders, time, and culture.
- *TBT Insight:* Connection amplifies energy, creating networks that sustain motion, alignment, and growth.

Love as the Foundation of Connection

Jesus taught that love is the force that binds all connections, aligning individuals with God, each other, and the universal blueprint.

- *Scripture: "Love the Lord your God with all your heart... and love your neighbor as yourself."* (Matthew 22:37-39)

- *TBT Insight:* Love is the energy that sustains connection. It aligns systems with purpose, fostering harmony and growth.

TBT Insight: Jesus demonstrated connection as the web that unites all systems—God, humanity, and creation. Through love and alignment, systems achieve balance, resilience, and purpose.

Summary

In His life and teachings, Jesus embodied the universal principles of the *Transcendent Blueprint Theory*:

- **Motion**: Purposeful action created momentum and disrupted stagnation.
- **Cycles**: Death gave way to life, and renewal followed endings.
- **Reciprocity**: Giving without expectation restored balance and abundance.
- **Transformation**: Lives and systems were reshaped, aligning with truth and purpose.
- **Connection**: Love united individuals, systems, and creation into a thriving whole.

And now back to us....

Chapter 10: Humanity — An Advanced Look at the Living Blueprint

1. Introduction: Humanity as the Nexus of Universal Laws

Humanity stands at the crossroads of universal laws, simultaneously a participant in and a shaper of the systems that define existence. Governed by the same principles that shape nature, thought, and energy, humanity reflects the dynamic interplay of cycles, reciprocity, transformation, motion, and connection.

Now I know I have veered slightly off course with my initial plans for this section of the book. But humanity is unique. Unlike trees rooted in one place or stars bound by gravity alone, humans possess the capacity for conscious action. This ability to choose—to align with or disrupt the universal blueprint—grants humanity extraordinary power but also profound responsibility.

Through the lens of the *Transcendent Blueprint Theory* (TBT), this chapter examines humanity as a living blueprint: a dynamic system governed by universal laws, while capable of shaping its own destiny, still bound by those laws.

2. The Cycles of Human Existence: Rhythms That Define Us

Human existence unfolds in cycles, echoing the rhythms of nature, the cosmos, and thought. These patterns are not isolated; they are deeply embedded in the universal principles of cycles, transformation, and balance. By examining biological, historical, and personal cycles, we uncover the forces that shape our lives and humanity as a whole.

Biological Cycles: Connection to the Larger Ecosystem

At the core of human existence are biological cycles that sustain life, such as the circadian rhythm, reproduction, aging, and renewal. These cycles are deeply tied to the natural world, reminding us that humanity is part of a larger ecosystem governed by universal laws.

- **Circadian Rhythms and Ecosystem Flow**: The daily rhythm of sleep and wakefulness mirrors the cycles of activity and rest in ecosystems. Just as predators and prey alternate periods of action and repose, humans thrive when their biological rhythms align with natural light and darkness.

> - *Example:* Disruptions to circadian rhythms, such as shift work or excessive screen time, mirror ecological imbalances. Just as deforestation disrupts the flow of an ecosystem, chronic misalignment with biological cycles leads to stress, disease, and systemic inefficiency.

- **Cycles of Renewal in Biology**: Cellular renewal and reproduction reflect the universal principle of transformation. For instance, the body constantly regenerates itself, shedding old cells to make way for new growth, just as a forest renews itself through cycles of decay and regrowth.

Biological cycles illustrate humanity's deep integration with the rhythms of the natural world. By aligning with these patterns, we sustain health and vitality, reinforcing the universal blueprint that governs all living systems.

Historical Cycles: Civilizations as Ecosystems

History demonstrates that civilizations, like ecosystems, rise, flourish, decline, and renew. These patterns are neither arbitrary nor chaotic; they are governed by the same principles of cycles and transformation seen in nature.

- **Rise and Renewal**: Civilizations often grow during periods of abundance and innovation, only to face decline due to imbalance, overuse of resources, or external pressures. Yet from these declines often emerge periods of renewal and transformation.

> - *Example:* The Renaissance, which brought extraordinary advances in art, science, and philosophy, emerged from the perceived darkness of the Middle Ages. Much like a forest regenerates after a wildfire, humanity finds new growth in the ashes of past challenges.

- **Parallels with Ecosystems**: The fall of ancient empires, such as Rome, mirrors what occurs in overexploited ecosystems. Both demonstrate that imbalance leads to collapse, but renewal often follows as systems realign with sustainable principles.

Historical cycles remind us that even decline is part of the process of renewal. By studying these patterns, humanity can align itself with the rhythms of history, fostering resilience and adaptability in the face of change.

Personal Cycles: Growth, Struggle, Renewal

On an individual level, humans experience cycles of growth, struggle, and renewal. These personal rhythms are deeply connected to the larger patterns of life and echo the universal principles of motion, transformation, and balance.

- **Growth Through Struggle**: Struggles, whether physical, emotional, or intellectual, are often precursors to transformation. Much like a tree strengthens its roots during a storm, humans build resilience through adversity.
  - *Example:* A career setback or personal loss may feel like an ending but often becomes the foundation for new growth, offering opportunities for reinvention and insight.
  - **Seasons of Life**: Personal experiences align with the metaphor of seasons: spring (growth), summer (flourishing), autumn (change), and winter (rest and reflection).
- **Cycles of Renewal**: Just as ecosystems recover after disruption, individuals find renewal in periods of rest, reflection, and reorientation. These phases ensure that growth continues, even after periods of stagnation or hardship.

By embracing the natural cycles of personal life, we learn to flow with the rhythms of growth and renewal, aligning ourselves with the universal blueprint that sustains all systems.

The Universal Thread: Cycles Connect Us All

Humanity's cycles—biological, historical, and personal—are not isolated phenomena. They reflect the same rhythms that govern the cosmos, ecosystems, and thought. These cycles demonstrate that growth is not linear but rhythmic, ensuring that transformation and renewal are always possible.

Key Insight: Aligning with Life's Rhythms

Understanding the cycles of human existence allows us to see struggle as a necessary precursor to renewal, decline as an opportunity for transformation, and rest as a vital part of growth. By aligning with these rhythms, rather than resisting them, we cultivate resilience, adaptability, and balance—both individually and collectively.

3. Reciprocity: The Dynamic Exchange of Energy

Reciprocity is the lifeblood of human systems, where energy, resources, and emotions flow in a dynamic exchange that sustains relationships, communities, and the planet. Through the lens of the *Transcendent Blueprint Theory* (TBT), reciprocity emerges as a universal principle that transcends biology and physics, governing the equilibrium of human existence.

Relationships as Exchanges: The Flow of Energy

Every interaction between individuals—whether through words, actions, or emotions—represents a reciprocal exchange of energy. These exchanges either nurture or deplete the systems they inhabit, reinforcing the principle that no relationship exists in isolation.

- **Acts of Kindness as Energy Cycles**: Kindness and generosity create positive ripples, reinforcing trust and connection within a relationship.
    - *Example:* A simple gesture, such as helping a neighbor, can spark gratitude and inspire them to pass on the goodwill, creating a feedback loop of positivity.
- **Parallel to Physics**: This mirrors Newton's Third Law, where every action has an equal and opposite reaction. In relationships, every positive action prompts a response, perpetuating cycles of growth and connection.
- **Selfishness and Exploitation as Energy Drains**: Conversely, relationships characterized by selfishness or exploitation disrupt reciprocity, draining energy and fostering imbalance.
    - *Example:* A one-sided friendship where one person continually takes without giving back leads to resentment, much like an unbalanced system in physics teeters toward collapse.

Human relationships thrive when reciprocity is balanced, demonstrating that the cycles of giving and receiving are essential for harmony and growth.

Global Reciprocity: Humanity's Stewardship of the Planet

Reciprocity between humanity and the planet is foundational to sustaining life systems. Through the *Transcendent Blueprint Theory*, we can see how the natural flow of energy and resources mirrors the cycles of give and take found across all domains. Humanity's actions—when aligned with these cycles—can maintain balance, but disruptions to reciprocity lead to balance changes in the system.

The Carbon Cycle as a Reciprocal System

The carbon cycle exemplifies nature's most profound expression of reciprocity, demonstrating the dynamic exchange of energy and resources across plants, animals, soil, and the broader environment. Carbon is not the problem; *it is the essential medium through* which energy flows, connecting and sustaining life. Fluctuations in carbon levels are not inherently problematic but rather reflections of a system constantly adapting to maintain balance—a process that is deeply resilient and not yet fully understood.

TBT Perspective

Carbon serves as the thread that binds energy cycles together, enabling transformation and connection across life systems. Disruptions, such as deforestation or changes in land use, do not "break" the cycle but instead prompt natural adjustments. These shifts in carbon reservoirs—from plants to soil, oceans, or the atmosphere—are not failures but adaptive responses, revealing the inherent resilience and interconnectedness of the system.

- *Example:* **Forest Ecosystems and Carbon Reciprocity** Forest ecosystems offer a perfect illustration of carbon reciprocity. Through photosynthesis, trees absorb carbon dioxide, storing energy in their tissues while releasing oxygen. This process highlights the seamless balance of energy within natural systems. When forests are removed, the system dynamically redistributes carbon to other reservoirs, such as the atmosphere or oceans. This is not a sign of collapse but an adaptive rebalancing, showcasing the system's ability to respond to change.

The carbon cycle demonstrates the adaptability and self-correcting nature of systems, as outlined by the *Transcendent Blueprint Theory*. Fluctuations in carbon reflect systemic adjustments rather than inherent problems. This perspective encourages us to move away from interventionist approaches and instead focus on understanding the dynamic reciprocity at play. By observing these patterns, we deepen our appreciation for how natural systems achieve balance, even amid disruption.

Furthermore: Carbon as Balance, Not Enemy

Carbon is not the enemy; it is the balance. Its fluctuations are not failures but testaments to the system's dynamic adaptability. Through the lens of the *Transcendent Blueprint Theory*, carbon becomes a vital conduit for energy transfer and a cornerstone of life's interconnected systems. By recognizing it as part of life's cycles rather than a problem to be solved, we can better understand how natural systems achieve equilibrium.

TBT Parallel

The carbon cycle reflects a universal truth about the flow of energy, both in natural ecosystems and human systems. Stagnation or imbalance arises when resources, energy, or ideas are hoarded or withheld. Carbon's dynamic movement mirrors this principle: life is sustained through the constant exchange and renewal of resources. Just as the release of oxygen sustains ecosystems, the natural sequestration of carbon ensures equilibrium, underscoring the importance of reciprocity and balance across all systems. (This is an awfully big mention of the subject, which is important to most, and due to this fact, I will dedicate a whole chapter later in the book)

Human Impact on Planetary Reciprocity: A Natural Alignment

Through the lens of the Transcendent Blueprint Theory, humanity is not a manager of Earth's systems but a participant in its greater complexity, governed by the same unchanging universal laws. The challenge is not to "fix" or "control" perceived imbalances but to realign our actions with the natural principles already written into the fabric of existence.

TBT Perspective

Carbon is not a pollutant to be managed—it is a vital component of the Earth's cycles, a cornerstone of energy exchange and life itself. Human activity, like deforestation or energy use, may trigger shifts in the system, but these shifts are not signs of failure. Instead, they demonstrate the system's profound capacity to adapt and maintain balance. The belief that we can "capture" or "control" carbon ignores the TBT principle that every action must abide by the same immutable laws, regardless of intention.

Example (Transcendent Lens)
When forests are removed, the system dynamically adjusts by redistributing carbon to other reservoirs, such as soil and the atmosphere. This adaptation is not a flaw but a signal of the system's inherent resilience. The human role, as revealed by TBT, is not to engineer solutions but to harmonize with the natural flow—by understanding the reciprocal nature of energy and resource exchange and participating within its cycles rather than against them.

The universal laws of the *Transcendent Blueprint Theory* reveal that humanity, like all systems, is bound by the same principles of cycles, reciprocity, and balance. Efforts to manage or control these systems often reflect a misunderstanding of their inherent complexity. Real alignment comes not from imposing human solutions but from recognizing and honoring the universal laws that govern all existence. By living within these rhythms, humanity fulfills its role as a participant in, not a manipulator of, the interconnected web of life.

Social Reciprocity: Misaligned Systems and the Laws of Adjustment
While reciprocity is a cornerstone of natural and human systems, it must flow organically to sustain balance. Attempts to impose forced equality—such as the redistribution of wealth under the hope of reciprocity—often conflict with the universal principles outlined in the *Transcendent Blueprint Theory*. These disruptions do not correct imbalances; they create new dynamics that trigger systemic responses under other universal laws.

Misinterpretations of Reciprocity
Efforts to enforce equity often mistake equality for balance, misunderstanding how reciprocity operates in nature and society:

- **Natural Reciprocity Is Dynamic, Not Equal**

In ecosystems, reciprocity reflects the flow of energy and resources, not rigid equality. Each participant contributes and receives based on its role and capacity, creating resilience through diversity and specialization.

- *Example:* A forest operates on reciprocal exchanges between trees, fungi, and animals, but these exchanges are not "fair" in a human sense. A large oak may absorb more sunlight than smaller saplings, but it provides shelter, food, and stability that benefit the whole system.

- **Imposed Equality Disrupts Cycles**

Forcing uniformity or redistribution where it does not naturally flow can create new imbalances. Redistribution of perceived equity often acts as a "sink," draining energy or resources from one area and unintentionally creating opportunities for a new "source" to emerge. While the intention may be fair and equal distribution, the result often diverges, leading to unintended consequences and reinforcing the dynamics of imbalance.

Reciprocity is not about achieving perfect balance; it is a dynamic interplay—an ebb and flow, a give and take, and a pendulum that swings just outside equilibrium. This movement reflects the adaptability and resilience of systems rather than rigid equality, demonstrating that balance is achieved not through uniformity but through the natural rhythms of exchange and transformation.

- **Example (TBT Lens):** In economies, forced wealth redistribution can weaken the productive forces that naturally generate abundance, creating a void that is often exploited by monopolies or opportunistic entities. This mirrors disrupted ecosystems, where imbalances create opportunities for invasive species to dominate and destabilize the environment further.

TBT Parallel: The Role of Sinks and Sources in Redistribution

As highlighted by the TBT, artificial disruptions to natural cycles—such as forced redistribution—do not achieve the intended balance but instead activate a secondary law: the creation of sinks and sources. This dynamic demonstrates how systems adjust to maintain motion and equilibrium:

- **Sinks**: Redistribution often weakens the existing flow of energy or resources, creating a vacuum where balance cannot be sustained.
  - *Example:* The collapse of an economic class or system creates a power void, allowing new entities to

consolidate influence and perpetuate inequality under different structures.

- **Sources**: As the vacuum created by the sink draws new energy, it empowers entities with adaptability or strategic advantage, realigning the system.
  - *Example:* Historical wealth shifts have frequently led to the rise of new elite classes, illustrating how transformation aligns with the laws of motion and reciprocity.

Forced interventions in natural systems, whether ecological or societal, do not halt the cycles of balance and imbalance but shift them into new configurations. The emergence of sinks and sources underscores the TBT's principle that energy and resources will always seek equilibrium, even if that equilibrium is temporary or destabilizing.

Pushing the Wheels of Change

Social Justice, while often motivated by perceptions that bring about feelings of unfairness or imbalance, may unintentionally serve as agents of systemic transformation. Their actions, driven by inequities, activate other universal laws, particularly:

- **Motion and Friction**: Their efforts generate resistance and counter-responses, refining systems through tension.
- **Contagion**: Their ideas spread, sparking societal ripples that initiate broader cycles of change.
- **Transformation**: By disrupting existing systems, they catalyze adaptations that create new balances—though not always the ones they intended.

Key Insight: Alignment, Not Imposition

The Transcendent Blueprint Theory reveals that reciprocity cannot be imposed; it must flow in alignment with the roles and capacities of participants within a system. Efforts to force balance disrupt natural dynamics, triggering new cycles of sinks and sources. True balance arises not from human intervention but from systems adjusting to align with universal laws.

Reframing Reciprocity as Universal

Reciprocity, through the lens of the *Transcendent Blueprint Theory* (TBT), transcends specific systems, reflecting a universal principle of balance and exchange. It is the rhythm by which giving and receiving sustain life—whether in ecosystems, societies, or personal relationships.

- **Parallel Cycles in Nature and Society**:

Just as ecosystems thrive on the exchange of resources, such as oxygen and carbon dioxide between plants and animals, societies flourish when contributions and benefits are in harmony. Reciprocity ensures the survival and renewal of systems, whether ecological or cultural.

> - *Example:* Cultural exchange, where knowledge and traditions are shared, strengthens relationships and fosters mutual respect, much like nutrient cycling sustains ecosystems.

- **The Strain of Overextension**:

In any system—natural or social—overexploitation or imbalance creates strain, disrupting harmony.

> - *Example:* Over-harvesting in nature depletes resources, much as overburdening individuals or communities erodes social cohesion. Both scenarios highlight the need for equitable cycles to sustain health and vitality.

When viewed through the TBT, reciprocity is not merely an ethical ideal but a universal law governing all systems. Its principles apply seamlessly across disciplines, demonstrating that balanced exchanges—whether in nature, society, or thought—are essential for stability and growth. By embracing reciprocity, humanity can align with the rhythms of the universe and foster systems that reflect harmony rather than discord.

The Universal Thread of Reciprocity

Reciprocity is the connective tissue of humanity, operating at every level—personal, communal, and global. It reflects the universal principle that systems thrive when energy flows freely, fostering balance and renewal. Disruptions to reciprocity—whether in relationships, ecosystems, or societies—do not spell failure but signal an opportunity for rebalancing and growth.

> - **TBT in Action**: Reciprocity is a universal thread that transcends disciplines, linking human behavior to the same cycles, exchanges, and feedback loops that govern the natural world.

For example, relationships flourish when balanced, just as ecosystems thrive on nutrient exchange.

Aligning with reciprocity allows humanity to create self-sustaining systems that reflect the rhythms of the universe, ensuring harmony within relationships, communities, and the planet.

Transformation: Humanity Tethered to the Universal Laws

Humanity's transformative capacity is often mistaken for a limitless ability to reshape the world, but this notion fails to account for the universal laws that tether all systems, including humans. While humanity can appear to act outside these laws—through technology, innovation, or even the dream of colonizing Mars—its actions are always subject to the boundaries set by cycles, reciprocity, transformation, and balance.

Humanity's Perceived Freedom vs. True Constraints

Humanity's creative potential is often celebrated as the ability to transcend natural systems. However, this is a misconception. Humans, like fungi or gravity, operate within the constraints of all universal laws. The difference lies in humanity's cognitive ability to make conscious choices, sometimes pushing the boundaries.

- **Fungi** *Example:* Fungi decompose organic material, instinctively fueling cycles of renewal and maintaining balance within ecosystems. They operate seamlessly within the natural blueprint, perpetuating the flow of energy and resources. Humans, however, possess the cognitive ability to disrupt these cycles through over-extraction or exploitation of resources. While fungi adjust to their environment without disrupting systemic equilibrium, human actions often push systems toward imbalance. Yet even when humans create instability, they remain tethered to the universal laws of balance and reciprocity, which inevitably force correction—whether through adaptation, innovation, or decline. For instance, when resource extraction surpasses sustainable limits, ecosystems begin to collapse, compelling humans to adjust their practices or face systemic failure.

The Mars Idea: A Case Study in Tethering

The idea of living sustainably on Mars illustrates humanity's overestimation of its freedom from universal laws. Mars lacks the atmospheric, gravitational, and ecological systems that sustain life on Earth. While humans may temporarily survive with artificial interventions, they remain tethered to Earth's natural cycles—dependent on oxygen, water, and energy systems that replicate terrestrial conditions. This dream of escaping Earth reflects a temporary misunderstanding by Humanity of the *Transcendent Blueprint Theory*: humanity can innovate, but it cannot transcend the laws that govern all systems.

Humanity's tethering to the universal blueprint highlights its dual nature as a system within and subject to natural laws. While humans can act creatively and disruptively, they remain bound to the cycles and principles that sustain life, proving that no system—no matter how advanced—can exist outside the Transcendent Blueprint Theory. *I hope in this one statement, understanding to humanity, of the gravity of the Theory.*

Motion and Inertia: Humanity in Action

Humanity's journey is driven by the universal principles of motion, resistance, and inertia, which govern cycles of momentum, stagnation, and renewal. Through intention and energy, human actions either align with or disrupt these cycles, reflecting the broader dynamics of the *Transcendent Blueprint Theory*.

The Momentum of Action

Intentional and aligned motion creates ripple effects, amplifying energy and driving progress. However, this momentum is most effective when it resonates with natural rhythms and universal laws.

> - *Example:* The collective energy of movements like the civil rights movement demonstrates the power of sustained, aligned motion to reshape societal structures. This parallels natural phenomena such as rivers carving paths through resistant terrain—not through confrontation but persistence and alignment with gravity, illustrating how cumulative effort, rather than singular force, drives transformation.

Excellent point—decline isn't just about material imbalance but also the stifling of free thought, free speech, and the open exchange of ideas.

When systems suppress intellectual and cultural reciprocity, they create conditions for stagnation and eventual collapse. Rome's decline wasn't solely about wealth hoarding; it was also about the erosion of societal structures that once fostered innovation, freedom, and civic engagement.

Inertia and Decline: Stifling the Flow of Humanity

When human systems lose motion or suppress cycles of reciprocity—whether in resources, ideas, or freedoms—they risk suffocating progress and falling into decline. Decline is not merely the result of hoarding wealth or neglecting systems; it stems from the breakdown of free exchange—of resources, thoughts, and voices—that sustains growth.

- **The Role of Thought and Speech**: Free thought and speech are the lifeblood of societal reciprocity. When these freedoms are suppressed, systems become rigid, unable to adapt or renew.
  - *Example:* Rome's decline was not simply due to wealth hoarding but also the loss of civic participation and intellectual freedom. As autocratic rule replaced the republic's collaborative systems, the stifling of ideas and public discourse hastened societal stagnation.

- **Acts Against Reciprocity**: Suppressing free thought and speech disrupts the cycles of creativity and progress, much like blocking a river halts its flow. Societies that restrict the exchange of ideas create "sinks" where innovation and trust are drained.
  - *Example:* Modern examples of censorship and authoritarian control show how restricting speech and dissent leads to social fragmentation, economic instability, and cultural decline.

- **Reciprocity and Renewal**: Societies thrive when they allow the free flow of resources, ideas, and voices. Aligning with the TBT principle of cycles ensures that systems continually renew themselves through open exchange and adaptive transformation.
  - *Example:* The Renaissance flourished because of the rediscovery and open exchange of knowledge, breaking through the intellectual stagnation of the Dark Ages.

Humanity thrives when systems of reciprocity—both material and intellectual—are upheld. **Reciprocity is the free movement of resources, ideas, energy, and interactions—not controlled allocation.** It thrives on natural flow and balance, where systems allow themselves the natural cycles of giving and receiving, aligning with the universal principles aligned by the *Transcendent Blueprint Theory*.

Controlled allocation, on the other hand, often disrupts this organic balance, introducing artificial structures that hinder adaptability and can lead to stagnation or decline. Natural reciprocity flourishes in environments where freedom, trust, and intention guide exchange rather than imposed restrictions or forced redistribution.

Decline occurs not so much from imbalances in wealth, but the prevention of natural fluctuations in wealth, the suppression of free thought and speech, which disrupts the cycles driving progress. By embracing these principles, humanity can avoid stagnation and foster enduring growth and renewal.

The Push and Pull of Human Interaction

Human motion creates fields of influence, akin to magnetism, which attract or repel based on intention, alignment, and energy. This dynamic interplay shapes relationships, movements, and collective progress.

- *Example:* A calm and steady leader projects minimal motion, creating a balanced field that encourages gradual alignment but may lack the momentum to drive significant change. In contrast, a leader who operates with abrupt energy and bold actions generates more intense motion, creating a strong push-and-pull effect that can rapidly attract or repel others. This dynamic mirrors natural systems, such as celestial bodies where sudden shifts in energy create gravitational waves, or magnetic fields where rapid motion intensifies attraction and repulsion.

Human motion exemplifies the TBT principle that systems thrive on dynamic energy. When actions are intentional and aligned with universal rhythms, they drive growth and renewal. Conversely, resistance, inertia, or misalignment risks compounding stagnation and decline. This underscores the importance of understanding and acting within the cycles and principles that govern all systems.

Connection: Humanity's Web of Interdependence

Humanity's strength is its ability to form connections—across relationships, systems, and disciplines. These connections are a dynamic force, fostering resilience and adaptation while also exposing vulnerabilities when balance is disrupted.

- **Interpersonal Connections:**

Relationships are the foundation of human systems, facilitating emotional stability, collaboration, and mutual support.

  - *Example:* Communities that prioritize authentic connection and reciprocity are more resilient in crises, demonstrating how shared energy stabilizes systems, much like interconnected ecosystems.

- **Global Interconnection:**

Advancements in technology and communication have expanded humanity's reach, creating unprecedented opportunities for collaboration. However, these networks also amplify disparities, inefficiencies, and potential imbalances.

  - **Key Insight:** Connection is neutral—it gains significance through how it is nurtured and aligned. Misaligned connections can create systemic strain, while purposeful ones foster innovation and cohesion.

  - **Connection to the Planet:** Humanity is inherently tied to the Earth's systems. Recognizing this connection shifts the narrative from dominance to reciprocity, fostering actions that align with natural cycles.
    - *Example:* Many Indigenous cultures emphasize reciprocity and respect for nature, offering lessons on how alignment with natural rhythms sustains both ecosystems and human communities.

Connection reflects the TBT principle of interdependence. Whether among individuals, global systems, or the planet itself, balanced and intentional connections promote resilience and renewal, while misaligned connections risk fragmentation.

Humanity as the Living Blueprint
Viewed through the lens of the TBT, humanity embodies the universal principles that govern all systems—cycles, reciprocity, transformation, motion, and connection. Humanity's actions and systems do not operate outside these laws; rather, they are shaped by them, even as humans attempt to innovate and create within these boundaries.

- **Humanity's Dual Role:**
  - **Participant:** Humanity is subject to the universal laws that govern motion, balance, and transformation, just like any other system.
  - **Creator:** Through intentional action and innovation, humanity shapes systems, reflecting the creative capacity of universal principles.
- **Parallels Across Disciplines:**

Human systems mirror the dynamics seen in physics, biology, and thought, highlighting the interconnectedness of all things.

- *Example:* Human thought cycles and societal evolutions mirror the nutrient cycles in ecosystems or the orbital dynamics of celestial bodies, reflecting the same balance-seeking patterns.

Humanity is both a reflection of and a participant in the *Transcendent Blueprint Theory*. By recognizing its interdependence with natural systems and aligning actions with universal principles, humanity can thrive within the cycles and flows that sustain existence.

I understand the need to clarify and refocus, especially after such a dynamic chapter where certain topics might come across as too pointed or politically charged. Here's a proposed revision for the conclusion of the *Humanity* chapter that acknowledges this shift, resets the focus, and transitions effectively back to the universal principles and their transcendence:

Conclusion: Humanity's Alignment with the Universal Blueprint
Humanity's position within the universal framework of the *Transcendent Blueprint Theory* is both extraordinary and humbling. As participants in the natural order, humans are tethered to the same principles that govern all

systems. Yet, our capacity for creativity, innovation, and disruption places us in a unique role—not separate from the laws of existence, but deeply intertwined with them.

The lens of the TBT offers clarity in a world where narratives about humanity's relationship with the planet, society, and each other often conflict. These narratives, no matter what their focus, can sometimes cloud the truth: that all systems—natural, social, and individual—are bound by the same principles of cycles, reciprocity, motion, transformation, and connection.

A Note on Perspective

It's easy to get lost in the details of issues or diverge into discussions that feel far removed from the heart of this theory. For that, I offer a moment of reflection. This chapter sought to delve deeply into humanity's role within the universal blueprint, but in doing so, may have touched on topics that distract from the core purpose of the TBT: to reveal the transcendent principles that guide all systems, across all disciplines.

These topics are important, but through the TBT lens, they are not isolated issues. They are expressions of the greater truth—that the universal principles transcend individual challenges, offering a broader framework for understanding and alignment.

Refocusing on the Principles

The Transcendent Blueprint Theory reminds us that our strength lies in understanding, not controlling, these principles. Humanity's greatest potential is realized not by imposing order on the chaos of the world, but by aligning with the cycles and rhythms that already exist. By focusing on the principles that transcend biology, physics, thought, and humanity itself, we can find clarity, purpose, and renewal.

As we continue, we'll examine these foundational principles, exploring their transcendence across systems and their implications for understanding everything from the smallest particle to the broadest societal structures. Together, we'll realign our focus, letting these universal truths guide us toward a deeper appreciation of the blueprint that connects us all.

The last section will begin to explore the principles of the *Transcendent Blueprint Theory* in contexts that stretch our understanding—faith, miracles, religion, and transformation. And since we are taking on these hard subjects lets not forget about carbon. Like adaptation, this book is changing as I write it and we will see where it goes. These topics often evoke strong emotions, deep beliefs, or skepticism, but viewed through the TBT, they illuminate universal truths that bridge the tangible and intangible. I expect TBT to receive some initial pushback, but on the other side of that test will become a foundation of truth.

But before we embark on that exploration, I present to you the **X Chapters**. These chapters reveal groundbreaking insights and new theories that expand our understanding of the universe and life itself. Here, the **Transcendent Blueprint Theory (TBT)** comes to life, offering fresh hypotheses and profound revelations that challenge conventional thinking.

The X Chapters are more than ideas—they are a ride through the uncharted territories of thought, where the universal blueprint reveals its power in explaining the mysteries of existence. They set the foundation for the truths we will explore in part three, which is equally essential in completing this journey.

Prepare yourself for the revelation of the **X Chapters**—and enjoy the ride.

Part ∞
The X Chapters - Rethinking Thought

Chapter X: Neutrality and Freedom

"Stay in the right and you'll never lose a fight."

Words I have lived by—words that kept me safe, helped me defy death, and placed me in positions to help others. For years, I held onto them without fully understanding their depth. Only now do I see how these words connect to a universal law—a law that has shaped my life and operates in all things.

There is a magnetic force woven into the operating laws of the universe. Whether you like it or not, we are all subject to it. It governs energy, motion, and the way we navigate life.

It doesn't matter if you believe in creation or evolution—these rules apply to you.

My Early Lessons in Neutrality

This understanding began long before I had words for it. My earliest memories were formed in chaos, growing up with a schizophrenic mother whose actions created a turbulence I could feel even as a child. Though I was taken from that environment around the age of three, the lessons had already taken root. The storm she created taught me to detect energy, to sense the pull and push of forces around me.

I learned to be neutral—not as a choice but as a necessity. Confrontation wasn't just dangerous; it was like stepping into quicksand, dragging me into chaos. Neutrality became my shield, allowing me to observe, adapt, and move without getting pulled into the storm. It wasn't passivity; it was precision. It was survival.

Magnetic Forces and Universal Laws

In time, I realized neutrality wasn't just a personal refuge—it was a universal principle. Motion creates energy, and energy creates magnetic force. This isn't a metaphor; it's a physical law rooted in the dynamics of the universe. Motion attracts and repels, creating forces that shape everything they touch.

Neutrality, I've come to understand, is freedom. It is the ability to move without being consumed by these forces. It is the state where energy is conserved, potential is held, and balance is maintained. Without neutrality, you lose control, drawn into the currents of motion. With neutrality, you gain the ability to navigate forces, engaging them only when necessary.

Lessons from Neutrality and Freedom

Lesson 1: Freedom is a universal law, but it doesn't play by the same rules as force.

Freedom exists in neutrality, in the balance between extremes. It is the potential to act without being compelled. Just as a seed waits in dormancy, conserving its energy until the right conditions arise, neutrality holds the power of the unexpressed.

Lesson 2: Neutrality allows movement and discretion.

All those times I was in the right place at the right time—more lucky than good—weren't coincidences. People would tell me, "Thank God you were there," but I know now it was because I had learned to live in neutral. Neutrality allowed me to engage force at the right moment, without losing *control or being consumed* by it.

Lesson 3: Absent neutrality, freedom is lost.

Without a foundation of neutrality, you are pulled into the forces around you, losing control over your actions and direction. Neutrality is what keeps you grounded, giving you the ability to act with intention rather than being swept away.

Neutrality and Freedom as One Universal Law

Neutrality and freedom are not separate—they are the same universal principle, transcending disciplines and proving the validity of the **Transcendent Blueprint Theory (TBT).**

Neutrality is not the absence of force but its balance. In this balance, freedom resides, undiminished by extremes. Growth—the engagement of force—emerges from neutrality, drawing energy from its stored potential. This process reveals itself in all systems:

1. Neutrality as Freedom in Nature:

- A seed in dormancy holds its potential energy in a state of neutrality. This neutrality is its freedom—the capacity to grow under the right conditions. The moment the seed engages force, breaking dormancy and sprouting, it transitions into growth.

- Ecosystems achieve neutrality through balance, allowing species to thrive without over-consuming resources. When imbalance occurs, systems grow or shift to restore equilibrium.

2. **Neutrality as Freedom in Thought:**
   - Neutrality in thought provides freedom to hold ideas without bias, judgment, or premature action. It is the space where creativity thrives, unshaped by external pressures. Once a thought engages force—through action, debate, or decision—it grows but expends the freedom of its neutral state.

Growth as the Engagement of Force

Growth is the active engagement of force, emerging from the freedom inherent in neutrality. This principle is universal:

1. **In Nature:**
   - A tree's growth requires engaging forces—pushing against gravity, weathering storms, and competing for light. This process draws from the tree's neutral foundation, its stored potential as a seed.

2. **In Thought:**
   - Growth in thought occurs when ideas are tested, debated, and acted upon. This engagement requires effort, expending the neutrality of unchallenged potential.

Personal Reflections: Living in Neutral

Neutrality shaped my life, often without my understanding. It allowed me to avoid conflict, maintain balance, and move freely in moments when others were stuck. Neutrality wasn't weakness or avoidance—it was freedom.

Freedom, I've learned, is not the opposite of neutrality but its evolution. Neutrality gives you the foundation to act when necessary, to engage forces without losing control. It is the reason I've been able to help others, navigate chaos, and find balance in the face of extremes.

Conclusion: Neutrality and Freedom in the Transcendent Blueprint

Neutrality is freedom, operating under the same universal law. It is the transcendent principle that allows systems to hold potential, maintain balance, and grow. This principle applies equally to nature, thought, and the cosmos, proving the universality of the *Transcendent Blueprint Theory*.

From the stillness of a seed to the expansion of galaxies, from the clarity of thought to the engagement of action, neutrality and freedom are not distinct—they are the same truth. To understand this is to see the blueprint of existence: a cycle of rest, growth, and renewal, governed by the balance of forces and the freedom of neutrality.

These laws exist beyond us yet govern us. To understand them is to gain wisdom. And with wisdom comes the discretion to navigate the forces, rather than render control.

Chapter XX: Dark Matter and the Blueprint of Balance

The universe is full of mysteries, but none is more profound than **dark matter**. Invisible, elusive, and yet undeniably present, dark matter represents a force that shapes the cosmos while defying direct observation. For decades, scientists have struggled to define its true nature. Is it a particle? A modification of gravity? Or something entirely beyond our understanding?

Through the lens of the *Transcendent Blueprint Theory* **(TBT)**, dark matter emerges as more than a scientific puzzle. It becomes a testament to the **universal laws** that govern all things:

1. **Balance: The equilibrium required for systems to remain stable and thrive, seen in the way dark matter stabilizes galaxies and enables the universe to hold its form.**
2. **Energy Flow: The continuous movement and redistribution of energy that drives growth, motion, and transformation, reflected in dark matter's role in guiding the structure and motion of galaxies.**
3. **Neutrality as Potential**: The state of preserved potential energy and unexpressed force, evident in dark matter's invisible yet fundamental influence on cosmic stability and formation.

In the same way TBT unifies concepts across disciplines, dark matter reveals the deeper truths that link all systems—biological, physical, and cosmic—through these principles.

In this chapter, we explore dark matter as a manifestation of the **neutral universal law**, a principle of balance that transcends physics, biology, and thought.

The Mystery of Dark Matter

Dark matter is unlike anything we can see, touch, or measure directly. It does not emit, absorb, or reflect light, making it invisible to even the most advanced telescopes. Yet its presence is undeniable:

- **Galactic Stability**: Dark matter prevents galaxies from tearing apart despite their high rotational speeds. Without it, the visible matter in galaxies would be unable to hold together under the laws of gravity.

THE TRANSCENDENT BLUEPRINT

- **Cosmic Scaffolding**: Dark matter shapes the large-scale structure of the universe, forming a **cosmic web** where galaxies cluster like dewdrops on a spider's silk.
- **Mass-Energy Content**: Comprising approximately **27%** of the universe's total mass-energy content, dark matter vastly outweighs the ordinary matter we can observe.

Despite decades of research, dark matter's true nature remains elusive. Proposed theories include weakly interacting massive particles (WIMPs), axions, or even modifications to our understanding of gravity. But none have offered definitive answers.

The question of **what** dark matter is invites us to look beyond conventional explanations and into the universal truths of TBT.

Dark Matter as Balance

One of the core principles of TBT is **balance**: the idea that all systems, whether physical, biological, or conceptual, are guided by forces that seek equilibrium. Dark matter exemplifies this principle on a cosmic scale.

1. Galactic Stability:

- Dark matter acts as a stabilizer, balancing the gravitational pull of galaxies against the centrifugal force of their rotation. Without this invisible mass, galaxies would collapse or disintegrate.
- This reflects the **TBT truth of homeostasis**, where systems thrive when forces are balanced. Just as biological organisms regulate temperature, nutrients, and energy, galaxies rely on dark matter for structural integrity.

2. Cosmic Scaffolding:

- Dark matter forms a web-like framework across the universe, creating the gravitational "sinks" where visible matter gathers to form galaxies and clusters.
- This mirrors the **TBT truth of foundational balance**, where unseen forces provide the stability required for complexity to emerge. Much like the mycelial networks beneath forests support the growth of trees, dark matter underpins the growth of the universe.

Dark Matter and Source-to-Sink Dynamics

In biological systems, **source-to-sink dynamics** describe the movement of resources from areas of abundance (sources) to areas of need (sinks). Dark matter operates similarly, guiding the flow of matter and energy in the universe.

1. **Gravitational Sinks:**
    - Dark matter creates regions of high gravitational potential, attracting ordinary matter into these "sinks." Within these wells, galaxies form and evolve, much like plants grow in nutrient-rich soils.
    - This reflects the **TBT truth of energy flow**, where resources—whether nutrients or gravitational potential—move along gradients to sustain growth and structure.

2. **Cosmic Energy Flow:**
    - Dark matter influences the motion of galaxies, redistributing energy from dense regions to less dense ones, maintaining the balance of the cosmic web.
    - This dynamic aligns with the **TBT truth of reciprocity**, where systems exchange resources to sustain balance and growth.

The Cosmic Web: Dark Matter's Unifying Role

Dark matter's influence is not confined to individual galaxies but spans the universe, linking systems into a vast, structured network. This unifying role reflects core **TBT truths** of energy transfer and the seamless exchange of forces across boundaries.

1. **The Cosmic Framework:**
    - Dark matter forms vast filaments that act as pathways for the flow of matter and energy, guiding the arrangement of galaxies into clusters. These filaments serve as the gravitational backbone of the universe, allowing structure to emerge from apparent chaos.
    - This reflects the **TBT truth of universal scaffolding**, where unseen forces create the framework necessary for complexity to arise and sustain itself.

2. **Unseen Forces at Work:**
   - Although invisible, dark matter exerts a gravitational pull that shapes the motion and placement of galaxies, much like the unseen forces in biological systems guide growth and interaction. Just as wind moves trees or mycelial networks transfer nutrients, dark matter organizes the flow of cosmic energy.
   - This aligns with the **TBT truth of influence beyond visibility**, where systems rely on forces that, while unseen, shape every observable outcome.

Contextual Variability: Dark Matter's Role in Universal Balance

Another core TBT truth is **contextual variability**: the idea that universal laws manifest differently depending on scale, purpose, and environment. Dark matter exemplifies this principle.

1. **Galactic Scale:**
   - On the scale of individual galaxies, dark matter acts as a stabilizer, maintaining balance and preventing collapse. Its role is akin to a biological system regulating internal balance to sustain life.
2. **Cosmic Scale:**
   - Across the universe, dark matter organizes matter into structures, creating the scaffolding for growth and change. Its influence shifts from local stability to global formation, showing how the same principle adapts to different contexts.

Lessons from Dark Matter: A TBT Perspective

Dark matter offers profound lessons when viewed through the lens of **TBT**, demonstrating how universal principles apply across systems, from the smallest scales to the vast cosmos:

1. **Balance and Growth Are Intertwined:**
   - Dark matter provides the equilibrium galaxies need to hold their structure and grow. This reveals a universal principle: balance is the foundation of complexity. Without balance, growth becomes unsustainable—whether in ecosystems, societies, or personal development.

2. **The Power of the Unseen:**
    - Dark matter reminds us that the most transformative forces are often invisible. Its gravitational pull shapes the cosmos just as unseen energies, like thoughts or emotions, influence human behavior. This reflects the universal truth that true power often resides in the unseen.
3. **Connection Enables Stability and Expansion:**
    - The gravitational pathways created by dark matter allow galaxies to form and sustain themselves. This demonstrates a universal law of systems: stability and growth depend on pathways of exchange and support, whether they involve matter, energy, or ideas.

Conclusion: The Blueprint of the Cosmos

Dark matter challenges us to look beyond what is visible and measurable, inviting us to explore the deeper forces that shape the universe. Through the lens of **TBT**, dark matter is not merely a cosmic anomaly but a profound demonstration of universal truths.

It balances galaxies, guides energy flow, and creates the framework for systems to thrive. These principles—**balance, energy flow, reciprocity, and the transcendence of biological laws across disciplines**—are the same truths that govern life, thought, and motion at every scale.

By understanding dark matter through **TBT**, we recognize that the universe operates under shared principles, where patterns seen in biological systems extend to the cosmos. From the smallest cell to the farthest galaxy, we all function under the same laws—the transcendent blueprint unites the visible and invisible, the known and the unknown, into a single, cohesive reality. But lets not stop there.

A New Perspective on Dark Matter: A TBT Hypothesis

Through the lens of the **Transcendent Blueprint Theory**, dark matter is not simply an invisible mass or a particle to be discovered—it is a fundamental expression of universal principles. Its behavior transcends traditional physical models, offering profound parallels to biological systems and illuminating the hidden laws that govern all things. The following propositions offer a new framework for understanding dark matter:

THE TRANSCENDENT BLUEPRINT

1. **Dark Matter as Neutral Energy Stabilizer:**
   - Dark matter functions as a neutral force, preserving potential energy and maintaining equilibrium across cosmic scales. It acts as the unseen scaffolding that stabilizes galaxies and enables the universe to sustain its structure, mirroring the dormancy phase in biological systems where potential energy is conserved.

2. **Source-to-Sink Dynamics on a Cosmic Scale:**
   - Dark matter creates gravitational sinks that guide the flow of ordinary matter and energy. This dynamic mirrors biological systems, where resources move along gradients from areas of abundance to areas of need. Dark matter, therefore, acts not merely as mass but as a **gravitational guide** that facilitates efficient energy distribution and structural growth.

3. **Contextual Variability of Dark Matter:**
   - The influence of dark matter shifts based on scale:
   - At a **galactic scale**, it provides localized stability by counteracting centrifugal forces.
   - At a **cosmic scale**, it organizes matter into vast filaments, forming the framework for the universe's large-scale structure.
   - This adaptability reflects the TBT truth of **contextual variability**, where the same force manifests differently based on the system it governs.

4. **Dark Matter as an Archetype of Unseen Influence:**
   - Dark matter embodies the universal truth that unseen forces drive visible outcomes. Whether in the gravitational pull shaping galaxies or the subtle forces guiding human thought, dark matter's invisibility highlights the profound power of influence beyond direct observation.

5. **Testing the TBT Framework:**
   - By exploring parallels between dark matter and biological systems, such as nutrient flows or network dynamics, we can identify whether dark matter operates in patterned ways resembling life. For *example:*

- Does dark matter exhibit emergent properties similar to neural or mycelial networks?
- Can energy distribution influenced by dark matter be modeled using source-to-sink principles observed in ecosystems?

The Implications of a TBT Approach

This perspective redefines dark matter not as an anomaly to solve but as a **manifestation of universal laws**. It positions dark matter as:

- A stabilizing force that upholds balance across cosmic and galactic scales.
- A guide for energy and matter, reflecting biological principles of flow and reciprocity.
- A transcendent phenomenon that unites disciplines, showing that the same truths apply to both the cosmic and the biological.

By considering dark matter through the *Transcendent Blueprint Theory*, we open the door to new lines of inquiry that transcend traditional physics and bridge the gap between cosmic and living systems. This approach may not only help solve the mystery of dark matter but also provide a deeper understanding of the fundamental laws that shape existence.

Chapter XXX: The Seed of the Universe

From the smallest acorn to the grandest galaxy, the universe whispers a truth that transcends scale: everything begins with a seed. This chapter invites you to explore the universe itself as one such seed—a cosmic kernel born from potential, growing, dispersing, and perhaps, one day, planting anew.

The Seed Metaphor in the Transcendent Blueprint

A seed encapsulates possibility. Within its modest shell lies the blueprint for entire ecosystems, life cycles, and renewal. Yet, a seed is not merely a vessel of potential—it is a process, governed by the same universal laws that shape galaxies, trees, and human thought. It absorbs, expands, interacts, and eventually declines, leaving behind seeds of its own creation.

Viewed through the **Transcendent Blueprint Theory (TBT)**, the universe itself is a seed:

- **Galaxies as Offspring**: Scattered like windblown seeds across the cosmic soil, galaxies emerge as offspring of the universe, growing and sustaining the elements needed for life.
- **Stars as Energy Engines**: Stars, much like photosynthesis in trees, drive the universe's growth by transforming matter into energy, creating the elements that sustain existence.
- **Dark Energy as Sustenance**: The mysterious force driving the universe's expansion can be likened to the nutrient flow that supports growth in biological systems.

Yet within every seed lies the blueprint for its own decline. Just as a tree transitions from rapid growth to maintenance, the universe shows signs of shifting its focus. The seeds of entropy—the gradual dispersal of usable energy—are already embedded in its fabric. Over time, the universe, like a tree, may focus less on expansion and more on sustaining what already exists.

The Cosmic Tree: Dropping Seeds

Where does the seed of the universe come from? To imagine the universe as a seed implies the existence of a tree—something larger, older, and even more profound. While science cannot yet answer this question fully, several theories resonate with this metaphor:

1. **The Multiverse Hypothesis:**
   - The "tree" could be the multiverse, a vast system from which individual universes "drop" like seeds. Each universe, like each seed, carries its own potential, expanding and evolving according to its conditions.
2. **Black Holes as Seed Planters:**
   - Some physicists propose that black holes might be portals to new universes. Compressed matter and energy could form the seed of another cosmos, suggesting that black holes act as the mechanisms by which the multiverse propagates.
3. **Cosmic Cycles:**
   - The Big Bounce theory suggests that our universe is part of an infinite cycle of contractions and expansions. In this view, the seed of the universe is not a one-time event but a recurring pattern, echoing the natural cycles of growth, death, and renewal seen in forests and ecosystems.

Growth and Decline: The Tree's Lifecycle

Trees, like universes, do not grow indefinitely. As they age, their energy shifts from rapid expansion to sustaining their structure. The wood that supports their branches becomes both a strength and a drain on resources. This pattern mirrors what scientists observe in the cosmos:

1. **Star Formation Decline:**
   - The rate of star formation has decreased significantly since the universe's early days. The cosmic "forest" is no longer sprouting as many new stars.
2. **Dark Energy's Role:**
   - Dark energy, the force driving the universe's expansion, may one day slow its growth. Some theories even suggest the universe could reach a state of stasis or begin to contract, much like a tree entering its late stages of life.

This shift from growth to maintenance highlights a universal truth: no system, whether a tree or a universe, can expand forever. Energy must eventually be diverted to sustaining what already exists.

Is the Universe Full-Grown?
If the universe is a seed, is it fully grown? Or are we still witnessing its expansion phase? Observational evidence suggests we are somewhere in between:

- **Expansion Continues**: The universe is still growing, but not as explosively as during the beginning or the inflationary period.
- **Slower Star Formation**: The universe is transitioning, with growth slowing and maintenance becoming more prominent.

Like a tree in midlife, the universe balances growth and sustenance, preparing for an eventual decline.

Entropy and Renewal: The Next Seeds
If the universe is a seed, what happens when it reaches its end? Does it leave behind new seeds, or does its energy dissipate entirely? The answers may lie in two potential outcomes:

1. Heat Death:
- The universe could reach a state where energy is evenly distributed, and no work can be done. This ultimate state of entropy represents the decline of the cosmic tree.

2. Renewal Through Black Holes:
- Alternatively, black holes may act as mechanisms for renewal, compressing matter and energy into seeds for new universes. In this view, even in decline, the universe plants the seeds for future growth.

Every end carries the potential for a new beginning, much like a tree falling to enrich the soil for the next generation.

This idea of renewal is not unique to science. It aligns with the teachings of many religious traditions, which often emphasize cycles of death and rebirth, decay and regeneration. From ancient texts to modern interpretations, these spiritual perspectives echo the same universal truth: endings are not final—they are transitions to something new.

The TBT Truths Gleaned from the Seed Theory
Through the metaphor of the seed, **TBT** illuminates profound truths that apply to both the universe and life:

1. **Cycles of Growth and Decline:**
   - Growth is not infinite. Every system, from trees to galaxies, must eventually shift from expansion to maintenance, reflecting the balance required for sustainability.
2. **Balance as a Prerequisite for Complexity:**
   - The growth of the universe, like that of a tree, depends on balance—between energy flow, gravitational forces, and the sustaining influence of unseen elements like dark matter and dark energy.
3. **The Blueprint of Renewal:**
   - Even in decline, systems leave behind seeds for regeneration. This principle echoes across disciplines, from the reproductive cycles of plants to the potential for universes to arise from black holes.
4. **Transcendence of Scale:**
   - The seed metaphor reveals a transcendent truth: the same principles govern both the smallest biological systems and the vast expanse of the cosmos.

Conclusion — Before laying out our Hypothesis: The Seed as Universal Truth

The seed is not just a metaphor—it is a universal truth, bridging the biological and the cosmic. Whether we look at the sprouting of a sapling or the birth of a galaxy, we see the same principles at work: cycles of growth, maintenance, decline, and renewal.

To view the universe as a seed is to acknowledge its potential, its limitations, and its place in a larger system. It reminds us that everything, from the smallest cell to the vastest galaxy, is connected by shared principles—a transcendent blueprint that governs all existence.

The seed of the universe may not answer every question, but it offers a framework that connects the smallest act of growth to the vastest cosmic processes. In this way, the universe, like a seed, carries within it the promise of something greater, something beyond, something eternal.

The Seed-Cosmos Hypothesis: A New Framework for Understanding the Universe

Through the lens of the **Transcendent Blueprint Theory (TBT)**, the universe can be understood as a **seed**—a self-contained blueprint of potential, governed by universal principles that transcend scale. This hypothesis provides a unifying framework that connects biological processes and cosmic phenomena, offering insights into the nature of growth, decline, and renewal.

Core Propositions

1. **The Universe as a Seed:**

    - The universe operates as a seed, following a cycle of growth, maintenance, decline, and renewal:

    - **Growth Phase**: The beginning and cosmic inflation represent the universe's rapid germination, during which energy and matter expanded to establish the foundation for galaxies and stars.

    - **Maintenance Phase**: The present universe exhibits slower growth, with declining star formation and increasing focus on sustaining its complex structures.

    - **Decline and Renewal**: As entropy increases, the universe transitions toward stasis or renewal through mechanisms like black holes.

2. **Black Holes as Seed Dispersers:**

    - Black holes act as mechanisms of regeneration, compressing matter and energy into "seeds" that may form new universes. This aligns with the biological role of seeds dispersing potential for future growth.

3. **Renewal as a Universal Law:**

    - The cycle of death and rebirth, observed in ecosystems and spiritual teachings, extends to the cosmos:

    - The universe contains inherent mechanisms for renewal, whether through black holes, multiverse interactions, or cosmic cycles such as the Big Bounce.

4. **Biological Laws Transcend Scales:**

- The same principles that govern seeds, trees, and ecosystems—**growth, energy flow, balance, and renewal**—apply to the universe. This transcendence reveals the interconnected nature of all systems, biological and cosmic.

5. **Entropy and Renewal in Balance:**
   - The universe's eventual decline into maximum entropy (heat death) may not mark an end but a transition, where energy dispersal paves the way for new seeds of potential, mirroring the regenerative role of fallen trees in a forest.

Scientific Implications

The Seed-Cosmos Hypothesis invites exploration of several key questions:

1. **Black Hole Dynamics:**
   - Could black holes act as portals to new universes, dispersing seeds of energy and matter?
   - Are the patterns of black hole behavior analogous to biological cycles of reproduction and renewal?

2. **Cosmic Energy Flow:**
   - How do the principles of source-to-sink dynamics, seen in biological systems, manifest on a cosmic scale? Does dark energy function as the nutrient flow for universal growth?

3. **The Role of Cycles in the Cosmos:**
   - Is the universe part of a larger cycle, such as the multiverse or Big Bounce theories? How do these cycles align with TBT truths about growth and renewal?

The Seed-Cosmos Hypothesis in Context

This hypothesis bridges disciplines, connecting the biological and the cosmic:

- It aligns with spiritual teachings that emphasize cycles of rebirth and renewal, suggesting a universal truth underlying both scientific and philosophical traditions.
- It challenges linear models of the universe, reframing cosmic history as a cycle that mirrors the natural rhythms of life.

Conclusion

The **Seed-Cosmos Hypothesis** offers a bold reframing of the universe's origins, development, and destiny. By viewing the cosmos as a seed, it reveals the interconnected and cyclical nature of existence, bridging the gaps between biology, physics, and philosophy. Through this lens, we see the universe not as a static system but as a dynamic, evolving entity, carrying within it the promise of growth, renewal, and eternal cycles.

Chapter XL: Rethinking Time Through the Blueprint of Growth

Time is the great arbiter of existence, shaping our understanding of life, the cosmos, and everything in between. Even scripture reminds us: *"With the Lord a day is like a thousand years, and a thousand years are like a day"* (2 Peter 3:8). This ancient wisdom challenges our perception of time, suggesting that it may not flow linearly or uniformly as we assume but instead reflects the nature and context of the systems it governs.

The scientific consensus places the age of the universe at approximately 13.8 billion years, based on observations of cosmic radiation, light transmission, and the universe's expansion. But what if this timescale is not as rigid as we perceive? What if, as seen across **all natural systems—from the rhythms of ecosystems to the cycles of stars and the growth of trees—** the universe has expanded in phases: marked by rapid creation, stabilization, and eventual maintenance?

Through the lens of the Transcendent Blueprint Theory (TBT), the parallels between the universe and all systems governed by natural law reveal a deeper truth: time does not progress as a fixed measure but evolves in harmony with energy, growth, and complexity. Just as a tree begins with a seed—a blueprint encoded with the potential for its entire lifespan—so, too, did the universe emerge from a singularity, carrying within it the framework for galaxies, stars, and life.

From the birth of stars to the cycles of seasons, and from the vast scale of cosmic expansion to the intricate workings of cells, time flows within a rhythm that aligns creation, transformation, and balance. This perspective reframes time as more than a simple, linear march forward. Instead, it becomes a dynamic force—woven into the very fabric of existence—reflecting the phases of creation, growth, and renewal that we see echoed across the cosmos and within life itself.

1. The Tree and the Universe: A Shared Blueprint

As we learned the growth of a tree mirrors the expansion of the universe in profound ways:

1. **Seed to Fruition:**
    - A tree begins as a seed, a microcosm containing the potential for vast growth. Similarly, the universe may have started from a

"seed"—a singularity or act of creation—carrying the blueprint for galaxies, stars, and life itself.

2. **Rapid Early Growth:**
- In its youth, a tree grows rapidly, forming wide growth rings. This early phase mirrors the universe's **cosmic inflation**, where it expanded exponentially **in fractions of a second.**

3. **Slowing Expansion:**
- As a tree matures, its growth slows, and its rings become narrower. Likewise, the universe's expansion has decelerated over time, suggesting it has entered a phase of stabilization.

Key Insight: If we examined only the late, slower growth rings of a tree, we might wrongly assume its entire growth history followed the same pace. Applying this analogy to the universe challenges the assumption that its current state reflects its entire history.

2. The Scientific Basis for the Universe's Age

The estimate of 13.8 billion years comes from methods rooted in the observation of light and cosmic phenomena:

1. **Cosmic Microwave Background Radiation (CMB):**
   - The faint glow of radiation left over from the beginning provides a snapshot of the universe's early state. Scientists estimate its age by analyzing the CMB's temperature fluctuations.

2. **Hubble's Law and Expansion Rate:**
   - By measuring the rate of the universe's expansion (the Hubble Constant), scientists calculate how long it would take for the universe to reach its current size.

3. **Light Travel Time:**
   - The light from distant galaxies takes billions of years to reach Earth. By measuring this light's distance and speed, scientists approximate the universe's age.

While these methods are robust, they rely on the assumption of a **constant expansion rate** and an unchanging fabric of space-time—assumptions that TBT challenges.

3. Challenging the Timescale: Lessons from Trees

The tree analogy offers insights into how misinterpretations of growth phases might affect our understanding of time:

1. Wide Early Rings:
- A tree's early rings are wide because its energy is focused on rapid growth rather than maintenance. Similarly, the universe's early inflation phase accounted for most of its current size.

2. Narrow Late Rings:
- As a tree ages, its growth slows, producing narrower rings. The universe's slowing expansion could indicate a similar phase of maturity.

3. Misreading the Evidence:
- If we assume the universe's current expansion rate reflects its early phases, we risk overestimating its age—just as examining only the narrow rings of an older tree might misrepresent its rapid early growth.

4. Time and the Transmission of Light

Light is often considered the ultimate measure of time, but this approach assumes a constant and predictable framework. Through TBT, we see the potential for variability:

1. Constant Expansion:
- Current models assume the universe's expansion rate has remained predictable. However, like the wide growth rings of a young tree, the universe's early expansion may have been far more dynamic.

2. Uniform Space-Time:
- Light transmission relies on the assumption of a stable space-time fabric. Yet, dark matter, dark energy, or other forces may distort this fabric, making the universe appear older than it is.

5. Dark Matter: The Neutral Force

Dark matter, comprising about 27% of the universe, provides another clue. Its role mirrors the **neutral energy** in trees, stabilizing systems and shaping their growth:

1. **Stabilization:**
   - Dark matter acts as gravitational scaffolding, preventing galaxies from flying apart. This mirrors the structural maintenance of a tree as it diverts energy from growth to support its wood.

2. **Erosion or Collapse:**
   - As the universe's expansion slows, dark matter may play a role in its eventual transformation, echoing how trees succumb to rot or external forces in their later stages.

6. Implications of a Younger Universe

If the universe's age has been overestimated, it could reshape our understanding of cosmic history:

1. **Faster Early Growth:**
   - The universe's early expansion may have been far more rapid, compressing its timeline and creating the illusion of greater age.

2. **Dynamic Time:**
   - Time may not be a fixed, linear construct but a contextual variable, unfolding differently during periods of rapid growth versus stabilization.

3. **Revised Models:**
   - A younger universe would challenge existing assumptions about star formation, galaxy evolution, and the origins of life, prompting new questions about the nature of cosmic growth.

7. Transcendent Blueprint Truths

Through the TBT lens, the universe's relationship with time reveals deeper truths:

1. **Cycles and Transformation:**
   - Both trees and the universe grow in cycles, transitioning from rapid expansion to maintenance and eventual decline or renewal.

2. **Dynamic Time:**
   - Time, like growth, is not linear but contextual, varying based on the system's phase and conditions.

3. **Neutral Energy:**
   - Dark matter's role in stabilizing galaxies mirrors the neutral forces in trees that enable them to sustain their structures.
4. **Efficiency of Design:**
   - Whether in trees or the universe, the shift from rapid growth to stabilization reflects an underlying efficiency—a blueprint that optimizes energy use across all systems.

8. Conclusion: The Universe as a Living System

The universe, much like all living systems, grows in phases:

- **It begins with a seed**, expanding rapidly in its infancy before transitioning toward balance and stability.
- **Its current form tells only part of its story**, just as a tree's narrow, mature rings conceal the explosive energy of its youth.
- **Time itself is not a constant**, but a dynamic rhythm—a reflection of the system's growth, energy, and context.

To view the universe as younger or more dynamic than we've assumed does not diminish its grandeur; it magnifies it. It reveals a profound efficiency in its origin—a creation of elegant simplicity giving rise to unfathomable complexity. Through the *Transcendent Blueprint Theory* (TBT), we uncover a truth that resonates across trees, galaxies, and life itself:

Time is not a passive measure but an active participant in transformation—woven into the very fabric of existence. Like the rings of a tree or the unfolding of an idea, time evolves, mirroring the rhythms of growth, renewal, and purpose embedded within all systems.

In this light, the universe becomes not merely an expanse to be measured but a story to be understood—one that reflects the same universal laws guiding us, nature, and the cosmos toward growth, balance, and transformation.

This is the Contextual Time Hypothesis

Core Proposition:

Time is not a fixed, linear progression but a dynamic and contextual phenomenon. Its perceived flow and scale vary based on the system's state, phase, and conditions, reflecting universal principles of growth and transformation.

Key Principles:

1. **Dynamic Growth Phases:**
   - Time unfolds in phases of rapid expansion, stabilization, and decline, as seen in biological systems like trees and cosmic systems like the universe. Each phase has its own temporal characteristics:
   - **Rapid Early Growth**: Time appears to compress during early phases of rapid expansion, such as cosmic inflation or a tree's early growth rings.
   - **Stabilization and Maintenance**: Time seems to slow as systems stabilize, like the universe's current expansion or a tree's narrower late rings.
   - **Decline and Renewal**: Time may transition into cyclical renewal during phases of entropy or rebirth, such as the end of a star's life or the role of black holes in seeding new universes.

2. **Contextual Variability:**
   - Time is shaped by the **context of the system**:
   - Early universe phases operate under different principles (e.g., rapid inflation) than later phases of slower expansion.
   - Biological systems also demonstrate contextual time—e.g., dormancy in trees slows processes, while active growth phases accelerate them.

3. **Neutral Forces and Stabilization:**
   - Neutral forces, like dark matter in the universe or the structural stability of trees, contribute to the perception of time by maintaining equilibrium. These forces stabilize systems, influencing how time is experienced within them.

4. **Time as a Reflection of Energy Dynamics:**
   - Time is not an independent variable but a **function of energy flow and transformation**:
   - Early phases of high energy (cosmic inflation, rapid growth) make time appear compressed.
   - Later phases of declining energy (entropy, stabilization) create the illusion of slower, linear time.

Scientific Implications:

1. Revisiting Cosmic Timelines:
- The age of the universe may be overestimated if its early rapid growth phases compressed time. Current models based on light travel and expansion rates might require recalibration.

2. Dynamic Time Models:
- Models of cosmic history could integrate contextual variability, treating time as a system-dependent variable rather than a constant.

3. Parallel Studies in Biological Systems:
- The hypothesis invites interdisciplinary research, comparing time's contextual nature in trees, ecosystems, and cosmic systems.

Testing the Hypothesis:

1. Cosmic Evidence:
- Reanalyze cosmic microwave background (CMB) data to detect signs of **compressed time** during early inflationary phases.
- Explore whether dark matter's stabilizing role influences time perception across galactic scales.

2. Biological Parallels:
- Study tree growth patterns and how energy distribution during rapid growth mirrors time compression in cosmic phases.
- Investigate dormancy cycles in trees as analogs for stabilization phases in cosmic time.

3. Simulation Models:
- Develop computational models to simulate dynamic time across systems with varying energy flow, from trees to galaxies.

The Broader Implication:

The **Contextual Time Hypothesis** suggests that time is not an absolute constant but a **manifestation of growth, energy flow, and system dynamics**. By applying **TBT principles**, we see that time transcends disciplines, connecting the biological, cosmic, and conceptual into a unified framework.

Time and Evolution – A Shared Misunderstanding

If time itself is dynamic, contextual, and shaped by phases of growth and transformation, then our understanding of evolution—bound as it has been to a rigid interpretation of time—deserves careful reconsideration. Just as a tree's rapid early growth and its slower maturity reveal different temporal realities, the process of evolution may appear linear only because we have failed to account for time's variability.

By rewriting evolution through the lens of Contextual Time, we uncover a story not of gradual, uniform change but of bursts of transformation, stabilization, and renewal—patterns echoed in the universe, biological systems, and life itself. Time, as misunderstood in cosmic history, may have also obscured the true rhythm of what is not merely an evolutionary process, but instead bursts of rapid change followed by adaptation—patterns that align seamlessly with the principles of the *Transcendent Blueprint Theory*.

In the next chapter, we will explore how the *Transcendent Blueprint Theory* reframes evolution, revealing it as a dynamic process governed not by chance but by universal principles of energy, growth, and transformation—timeless laws woven into the fabric of existence.

Chapter L: Rethinking Evolution Through the Blueprint of Time

The theory of evolution, rooted in Darwin's observations and subsequent scientific expansions, suggests that all life evolved from common ancestors through mechanisms like natural selection and genetic mutation. While this perspective has shaped modern biology, the **Transcendent Blueprint Theory (TBT)** and the **Contextual Time Hypothesis** challenge its foundational assumptions, offering a broader framework to explain life's diversity and complexity.

TBT does not dismiss adaptation or genetic variation but reframes these processes within a universal blueprint that emphasizes **ordered principles**, **boundaries of change**, and the contextual unfolding of time.

1. The Lens of Contextual Time

Traditional evolutionary theory assumes a **consistent, linear timeline** over billions of years, during which species gradually transformed through random mutations and environmental pressures. However, the **Contextual Time Hypothesis** challenges this assumption, presenting time as **dynamic and variable**, shaped by energy and system phases.

1. Time Unfolds in Phases

Time progresses differently depending on the **growth phase** of a system:

- **Rapid Early Growth**: Systems experience bursts of rapid change when energy availability is high and environmental resistance is low.
- *Example:* A tree exhibits wide, rapid growth rings in its early years when resources are abundant.
- **Parallel in Life**: Early life forms may have undergone accelerated adaptations in high-energy conditions, resembling this early "growth phase."
- **Stabilization and Slower Change**: As systems mature, energy is redirected toward maintenance and equilibrium, slowing the pace of observable change.
- *Example:* Mature trees produce narrower rings, reflecting slower, stable growth over time.

- **Parallel in Life**: Life's later stages reflect similar stabilization, creating the perception of longer timescales for change.

This cyclical rhythm contrasts with the assumption of uniform, gradual transformation over billions of years.

2. Time Is Shaped by Energy Dynamics
The rate of change correlates with the **energy dynamics** of a system:

- **High Energy Phases**: In early life or the early universe, abundant energy enabled rapid development and bursts of transformation.
- **Cosmic Parallel**: The universe's inflationary phase occurred within fractions of a second, demonstrating rapid expansion under high energy conditions.
- **Low Energy Phases**: As energy diminished, systems entered stabilization phases where change slowed significantly.
- **Biological Parallel**: Life's apparent stasis in later ecosystems reflects this energy-driven deceleration.

Time, therefore, is not a fixed constant but unfolds **contextually**, adapting to the energy and maturity of the system it governs.

3. Scriptural Parallel: The Fall and the Changing of Kinds
The Bible offers a profound perspective on life's transformation and the introduction of change:

- **Original Harmony**: Before the Fall, creation existed in a state of **perfect balance and peace**, as God declared all things "very good" (Genesis 1:31). There was no death, suffering, or predation. Animals and humans coexisted peacefully, with plants given as food (Genesis 1:29–30).
- **Harmony of Design**: Life operated according to its original design, reflecting God's intention for creation—an ecosystem unmarked by decay or survival pressures.
- **The Fall as a Turning Point**: The entrance of sin brought about a **fundamental shift** in creation (Genesis 3:14–19):
- **Death and Suffering**: Death entered the world, breaking the harmony of the original creation (Romans 5:12).

- **Reordering of Relationships**: The curse disrupted the natural order. Animals that once lived peacefully (as implied by the pre-Fall context) now became predator and prey.
- **Cycles of Survival**: The natural world shifted into systems of survival, competition, and decay, introducing adaptation within the limits of original kinds.
- **Parallel to TBT**: The Fall can be viewed as a **"phase shift"** where creation was adjusted to operate under new conditions— **limited energy, survival pressures, and eventual entropy**. While life adapted to these challenges, its ability to change remained **bounded by its original blueprints**.
- *Example:* Animals now develop traits for survival—like sharper teeth or defensive behaviors—but remain constrained to their kinds.

4. Revelation: The Universal Law of Decay

The Fall of man not only shifted creation into a **phase of survival** but also instituted a **universal law** of decay and decline—one that governs all systems:

- **Cosmic Systems**: The universe's rapid inflation phase transitioned into a slower, stabilized expansion, now subject to the law of entropy (decline of usable energy).
- **Biological Systems**: Life entered cycles of aging, death, and renewal—adapting to the fallen world but always constrained by its original blueprints.
- **Ecological Systems**: Nature now balances growth, competition, and decay, operating under rhythms of survival and renewal.
- **Human Systems**: Human life, culture, and morality reflect patterns of growth, decline, and periodic renewal under the weight of this universal law.

This **universal law of decay** initiated at this time (fall) aligns perfectly with the *Transcendent Blueprint Theory*. TBT demonstrates that universal laws are not isolated to one system but apply across all of creation—biological, cosmic, and conceptual.

The **Fall of man** introduced a universal law of decay, shifting creation from its original phase of rapid growth and harmony into a long, extended phase of survival and decline. Life's adaptations within this new framework reflect **bounded change**—purposeful responses to survival pressures that remain constrained by their **original blueprints**.

This universal law now governs **all systems**, from the cosmos to biology. And in the context of Scripture, it reasons that **original design**—God's perfect creation—currently **coexists** with the modifications brought about by the **new law** of decay and survival, resulting from the Fall.

2. TBT and Evolutionary Similarities

Evolutionary theory often cites similarities between species as evidence of **common ancestry**. However, through the lens of the *Transcendent Blueprint Theory* **(TBT)**, these similarities are better understood as reflections of **shared universal principles** and design solutions, not descent from a single origin.

1. Shared Environment, Shared Rules

All life operates under the same **physical, chemical, and biological laws**, which shape form and function across species:

- **Gravity and Structure**: Skeletal systems in humans, monkeys, and other vertebrates are shaped to withstand gravity while enabling mobility. Even trees respond to gravity through their vascular structures, directing roots downward and shoots upward (gravitropism).

- **Energy Flow and Distribution**: Efficient energy distribution drives the formation of branching systems across kinds:

- **Blood vessels** in animals, **tree roots** in plants, and **river systems** in nature all reflect the same universal principle of minimizing resistance and maximizing flow.

- **Survival Constraints**: Similar environmental challenges—like mobility, resource distribution, or structural integrity—result in functional similarities across systems, independent of any evolutionary lineage.

These shared rules of existence produce recurring patterns, highlighting design consistency rather than ancestral relationships.

2. Common Design, Not Common Descent

Similarities between species arise from a **universal blueprint** that optimizes function, not from shared descent:

- **Branching Structures**: Branching systems appear repeatedly in trees, blood vessels, and neural networks because they are the most effective way to distribute resources efficiently across a system. This design reflects universal optimization, not evolutionary relationships.
- **Opposable Thumbs**: Humans, monkeys, and some other species possess opposable thumbs for grasping and manipulation. While similar in function, these adaptations arise from **common design principles** of mobility and precision, not shared ancestry.
- **Eyes Across Species**: Vertebrate eyes and cephalopod (octopus) eyes both enable vision but have fundamentally different structures. Their similarity reflects the universal need for sight, optimized independently within each kind.

These patterns of shared function demonstrate **purposeful design**, tailored to the needs of each kind, rather than a gradual transformation from one life form to another.

Key Insight

Similarities between species are best understood as **expressions of universal design principles** that respond to shared environmental and functional constraints. These patterns highlight a common blueprint for efficiency and survival—not evidence of common descent.

3. Adaptation vs. Transformation

The **Transcendent Blueprint Theory (TBT)** distinguishes between **adaptation within kinds** and the unsupported idea of one kind transforming into another. Adaptations optimize traits for survival or function but do not cross the inherent boundaries of a kind.

1. Adaptation Within Kinds

Species adapt to their environments through changes in traits such as size, behavior, or physiology. These changes remain within the boundaries of their kind:

THE TRANSCENDENT BLUEPRINT

- **Human Adaptations**: Humans exhibit adaptations to their environments that enhance survival while remaining distinctly human. For *example:*

- **Altitude Adaptation**: Populations living at high altitudes, such as those in the Himalayas or Andes, have developed greater lung capacity and increased red blood cell production to improve oxygen absorption. This adaptation allows humans to thrive in low-oxygen environments but does not alter their core identity as humans.

- **Selective Breeding**: Just as breeding in animals can amplify specific traits, humans have practiced selective breeding over centuries to influence traits like height, skin tone, or other physical characteristics. However, the variations remain firmly within the human kind.

2. Boundaries of Transformation

While species demonstrate flexibility to adapt, they cannot fundamentally transform into new kinds. Every kind is governed by a **distinct blueprint** that defines its range of possibilities:

- **Humans Are Fundamentally Distinct**: Humans and apes, despite superficial similarities, are markedly different in ways that extend beyond behavior:

- **Musculature and Skeletal Structure**: Human skeletons are optimized for upright walking, while apes are built for knuckle-walking or climbing. The pelvis, spine curvature, and foot structure in humans are uniquely designed for bipedal locomotion.

- **Skull and Brain**: The human skull houses a far larger and more complex brain, enabling higher cognitive functions like reasoning, creativity, and problem-solving—capacities far beyond any observed in apes.

- **Physiology and Skin**: Humans possess distinct sweat glands for temperature regulation, smooth skin with minimal hair, and a highly sensitive nervous system, reflecting a design suited for diverse environments.

- **Speech and Language**: Human vocal anatomy and neurological control over language allow for advanced communication unmatched in the animal kingdom.

These profound structural, physiological, and cognitive differences underscore that humans and apes belong to entirely separate kinds, governed by distinct blueprints.

4. Why Do Species Appear Similar?

The similarities observed between species are often misinterpreted as evidence of shared ancestry. The *Transcendent Blueprint Theory* **(TBT)** offers a different explanation: shared traits reflect **universal design principles** and biological efficiency, not direct lineage.

1. Universal Principles Drive Patterns

Shared structures across species are best understood as expressions of **efficient design solutions**, governed by universal laws:

- **Opposable Thumbs**: Both humans and certain primates have opposable thumbs, which reflect an optimized design for grasping and manipulation, not proof of common ancestry.
- **Branching Systems**: Patterns like branching in tree roots, blood vessels, and river networks emerge because they are the most efficient way to distribute resources and energy across a system.

The same principles apply across kinds because universal laws, like those governing energy flow and balance, transcend individual species.

2. Biological Efficiency

The physical and structural similarities between species arise from the **constraints of physics, chemistry, and biology**, reflecting functional efficiency:

- **Branching Patterns**: Trees, blood vessels, and rivers exhibit similar branching patterns because this structure minimizes resistance and maximizes flow, whether it's nutrients, blood, or water.
- **Skeletal Similarities**: Animals with limbs share structural similarities because their skeletons must contend with **gravity and locomotion**. For *example:*

- Birds, cats, and humans all have limb bones adapted for different functions (flying, running, or walking), but the underlying structure reflects the efficient response to physical forces.

These patterns highlight purposeful adaptations that optimize function within the constraints of natural laws, rather than transformation across kinds.

3. Misinterpreted Evidence

Genetic similarities between species are often cited as proof of common descent, but TBT reveals a different reality:

- **DNA Similarity**: Shared genetic sequences reflect the universality of biological mechanisms required to sustain life (e.g., protein synthesis, cellular respiration). These mechanisms are foundational across all living systems but do not indicate transformation from one species into another.

For example, humans and other mammals share significant portions of their DNA because they require similar systems for survival. However, the presence of shared DNA reflects a common design blueprint, not evolutionary ancestry.

Similarities between species are best understood as reflections of **universal design principles** and functional efficiency rather than evidence of transformation or shared ancestry. These patterns point to purposeful solutions governed by the same laws that shape all systems.

5. Rethinking Evolution's Timescale

The **Contextual Time Hypothesis** challenges the traditional view of evolution as a slow, linear process by showing that the rate of change is influenced by energy availability, system maturity, and environmental conditions.

1. Rapid Early Development

In the early stages of life—much like the early universe—high energy availability and minimal environmental resistance may have enabled **compressed bursts of change**:

- **High Energy Phases**: Early life systems experienced rapid development when conditions were ideal, similar to the exponential expansion seen during cosmic inflation.
- **Fossil Evidence**: The fossil record from early periods may reflect these **bursts of adaptation**, which appear sudden

because of their rapid nature, not gradual transformation over millions of years.

This challenges the assumption that significant change requires vast, linear timescales.

2. Stabilization and Adaptation

As ecosystems matured, energy flow stabilized, and systems entered phases of **maintenance and adaptation**:

- **Slower Change**: Once ecosystems achieved balance, the pace of adaptation naturally slowed, leading to the illusion of extended timelines for observable change.
- **Efficiency Over Expansion**: Mature systems focus on maintaining stability rather than dramatic shifts, aligning with patterns seen in later biological and cosmic development (e.g., a tree's slower growth in maturity).

This phase of stabilization creates a misleading impression that evolution has always progressed slowly and uniformly.

3. Dynamic Time

Change occurs in phases of **rapid expansion** and **stabilization**, reflecting the dynamic rhythms seen in biological and cosmic systems:

- **Early Life**: High-energy, rapid phases enabled bursts of adaptation and diversification.
- **Current Ecosystems**: Modern systems exhibit slower, more refined adaptations as they operate under stabilized conditions.
- **Parallel to the Universe**: Just as the universe expanded rapidly during its early phases and now grows steadily, life's development followed a similar rhythm of early bursts followed by stabilization.

The Contextual Time Hypothesis reveals that evolution's timeline is not uniform but **dynamic**. Rapid change in the early phases of life gave way to stabilization, creating the illusion of slow, gradual transformation. This rethinking aligns with the observed patterns of energy flow, adaptation, and system maturity across biological and cosmic systems.

6. TBT's Perspective on Kinds

The *Transcendent Blueprint Theory* **(TBT)** introduces the concept of "kinds" as the boundaries within which life evolves, emphasizing adaptation without transformation into entirely new forms.

1. Unique Blueprints

Each kind is governed by a **distinct blueprint**, which allows for flexibility in adaptation while remaining constrained by its core design:

- **Genetic Limits**: Life can express a range of traits within its kind—such as size, color, or behavior—without exceeding the boundaries of its encoded possibilities.
- **Dogs as an Example:** Through selective breeding, dogs have developed significant variation (e.g., from Chihuahuas to Great Danes), yet all remain within the canine kind.

These blueprints ensure stability and order across living systems, preventing transformation beyond their defined boundaries.

2. Shared Principles, Distinct Expressions

While kinds may share universal design principles, they express these principles in ways that maintain their **unique identity**:

- **Wings and Flight**: Birds, bats, and insects all exhibit wings for flight, but their structural and functional differences reflect their distinct blueprints within their kinds.
- **Limbs for Locomotion**: Mammals, reptiles, and amphibians may share limb structures that accommodate movement, but the expression of these traits remains unique to their kind.

Shared principles, such as efficiency and balance, transcend individual kinds, but these similarities do not imply shared ancestry.

3. Cycles of Growth and Renewal

Adaptations occur as part of the **cycles of growth, environmental interaction, and renewal**, but always within the boundaries of a kind:

- **Polar Bears and Grizzlies**: Polar bears have evolved thick fur, insulating fat layers, and white coats to survive Arctic conditions, while grizzlies have adapted to temperate forests and grasslands. Despite environmental adaptations, both remain within the bear kind.

- **Corn Varieties**: Humans have cultivated different varieties of corn (e.g., drought-resistant strains or high-yield hybrids), demonstrating adaptation within the plant kind through selective pressures.

These cycles reflect the dynamic relationship between life and its environment but do not represent pathways to transformation into new kinds.

TBT defines "kinds" as systems governed by **unique blueprints** that allow for adaptation within boundaries. Similarities across kinds reflect shared universal principles, while distinct expressions of these principles ensure that each kind retains its identity.

7. Why Aren't New Species Evolving?

A key critique of evolutionary theory is the absence of new life forms or kinds emerging today. The *Transcendent Blueprint Theory* **(TBT)** offers a compelling explanation, emphasizing stability, completed design, and the contextual nature of change.

1. The Seed of Life Is Complete

Life reproduces **within kinds**, reflecting the completeness of its original design:

- **Seeds and Trees**: A tree produces seeds that grow into other trees of the same kind. While variations occur—such as different leaf shapes or bark textures—these remain expressions of the tree's existing blueprint, not entirely new life forms.

- **Animals and Offspring**: Animals reproduce offspring that exhibit traits within the range of their kind. For example, lions and tigers may display subtle adaptations to their environments, but they remain members of the feline kind.

The diversity we observe today is a reflection of **variation within kinds**, not the emergence of entirely new blueprints.

2. Stability Over Chaos

Universal laws prioritize **stability** and balance to ensure ecosystems remain functional and ordered:

- **Interdependence**: Ecosystems rely on the stability of species relationships—such as predators, prey, and plant life—to maintain balance. The chaotic introduction of entirely new kinds would disrupt this finely tuned order.
- **Adaptation, Not Transformation**: While species adapt to changes in their environments, these adjustments occur within their existing boundaries. This ensures stability without introducing unpredictable, chaotic transformations.

3. Contextual Time and Stabilization

Change may have occurred rapidly during early phases of high energy but has **slowed as ecosystems matured and stabilized**:

- **Early Bursts of Change**: In the early phases of life, the energy conditions were ideal for rapid adaptation and diversification, aligning with the **Contextual Time Hypothesis**.
- **Current Ecosystems**: Today, ecosystems operate under stabilized conditions where significant adaptation is less frequent and dramatic, leading to the perception that evolution has stalled.

This slowing of observable change reflects the natural rhythms of growth and stabilization seen across biological and cosmic systems.

TBT explains the absence of new kinds as evidence of life's **completed design, governed by universal principles of stability and balance**. The adaptations we observe today occur within kinds, while the rate of change reflects the contextual nature of time and the stabilization of ecosystems.

8. Transcendent Blueprint Truths

The **Transcendent Blueprint Theory (TBT)** offers a unifying framework that challenges traditional evolutionary assumptions and reveals the patterns governing life's design and adaptation.

1. Cycles Over Linear Progress

Life develops through **cycles** of adaptation, stabilization, and renewal, rather than a continuous linear progression:

- **Adaptation**: Organisms respond dynamically to environmental pressures, optimizing traits for survival within their kind.

- **Stabilization**: Once systems mature, change slows, and energy shifts toward maintenance and balance, ensuring long-term stability.
- **Renewal**: When systems face decline or disruption, cycles of renewal emerge to restore balance and initiate new growth.

These cycles reflect the rhythms of life, aligning with universal principles of energy flow and transformation.

2. Shared Principles, Not Ancestry

Similarities between species are best understood as **expressions of shared design principles** that optimize function and efficiency:

- **Skeletal Structures**: Limb similarities across mammals reflect solutions to physical constraints like gravity and locomotion, not evidence of shared descent.
- **Biological Patterns**: Recurring systems like branching structures (e.g., veins, tree roots, rivers) arise because they are the most efficient way to distribute resources.

These patterns demonstrate purposeful design governed by universal laws, rather than randomness or ancestral relationships.

3. Boundaries of Change

Adaptation occurs within the **boundaries of kinds**, constrained by the blueprints that define each system's possibilities:

- **Bird Variability**: Birds adapt beak shapes, wing spans, or migration behaviors to suit their environments, but all remain birds.
- **Dog Breeds**: Selective breeding has produced incredible variation in size, behavior, and appearance within dogs, yet they remain within the canine kind.

The boundaries of change ensure stability and prevent transformation into entirely new forms, aligning life with universal design principles.

TBT reshapes evolutionary theory by demonstrating that life operates within **cyclical patterns of change**, reflecting shared design principles while maintaining boundaries defined by its blueprint.

9. Conclusion: Evolution Reframed Through the Lens of TBT

Through the *Transcendent Blueprint Theory* (TBT) and the Contextual Time Hypothesis, evolution is revealed not as a random, gradual transformation of one kind into another, but as a process governed by universal laws—phases of rapid change, stabilization, and adaptation.

This perspective challenges the traditional, linear view of evolution and reframes it as a reflection of the same dynamic principles that shape trees, galaxies, and the cosmos. Life's complexity is not a product of chance but of purpose-driven systems responding to energy, growth, and contextual rhythms.

By applying TBT, we uncover "evolution" as not really evolution as we know it but adaptation through a broader blueprint—one that transcends disciplines and reveals the uniformity of all existence, rooted in cycles of transformation and renewal.

The Transcendent Blueprint Theory of Change

Core Proposition

Life does not evolve by transforming from one kind into another; rather, it undergoes **bounded change**—a process of growth, adaptation, and transformation within **flexible yet distinct blueprints**. Governed by universal laws of balance, energy flow, and cycles, these changes unfold in **contextual time** rather than through linear progression.

Key Principles

1. Bounded Change

Life adapts within **predefined boundaries**, ensuring kinds remain distinct while maintaining flexibility to respond to environmental pressures:

- **Dogs and Selective Breeding**: The incredible variety in dog breeds (size, coat, behavior) reflects adaptation within the canine kind, but no amount of breeding transforms a dog into another kind of animal.

- **Human Adaptations**: Populations demonstrate physiological adaptations, such as increased red blood cell production at high altitudes or lactose tolerance in certain regions, yet all variations remain within the boundaries of the human kind.

These examples showcase the flexibility of life within the limits of its blueprint.

2. Contextual Time

Change occurs at **varying rates** depending on a system's energy and phase:

- **Rapid Early Growth**: In phases of high energy, systems undergo bursts of rapid change, similar to the early universe's inflationary period or a seed's sprouting phase.
- **Stabilization**: Mature systems slow their rate of change to focus on balance and maintenance, as seen in narrower growth rings in trees or ecosystems that have reached equilibrium.
- **Renewal or Decline**: As energy diminishes, systems transform, paving the way for renewal—such as a decaying tree providing nutrients for new growth.

This contextual approach reframes the rate of change as dynamic, not static or linear.

3. Universal Blueprint

Similarities across species and systems reflect **shared design principles** optimized for efficiency, not evidence of shared ancestry:

- **Branching Systems**: Found in blood vessels, tree canopies, and river networks, branching patterns optimize resource distribution.
- **Skeletal Structures**: Limb similarities across mammals reflect functional solutions to gravity and locomotion while remaining unique to each kind.

These recurring patterns are evidence of purposeful design responding to universal laws, not random evolutionary processes.

4. Cycles of Growth and Renewal

Change unfolds through **cycles** of adaptation, stabilization, and renewal, not through continuous, linear progression:

- **Growth**: Rapid adaptation or expansion occurs in early, high-energy phases, such as a sprouting seed or new ecosystems forming after a disturbance.

- **Stabilization**: Mature systems slow their pace of change, focusing on balance and sustainability.
- **Transformation**: End-of-life phases create conditions for renewal, as seen in the decomposition of plants and animals, which nourish the next generation.

These cycles reflect the natural rhythms of change embedded in all systems.

Testable Implications

1. Boundaries of Change
- Genetic studies should demonstrate variability within kinds (e.g., traits like size, color, or behavior) but no evidence of transformation between kinds.

2. Dynamic Rates of Change
- Fossil records should reveal bursts of rapid adaptation in early stages (similar to cosmic inflation) followed by slower, stabilized changes as systems mature.

3. Universal Design Efficiency
- Recurring patterns like branching structures or skeletal similarities should align with principles of **efficiency and resource distribution**, not common descent.

Scientific Context

The Transcendent Blueprint Theory of Change challenges traditional evolutionary theory by:

1. Replacing Randomness with Universal Laws
- Change is guided by ordered principles—balance, energy flow, and efficiency—not chance mutations or blind processes.

2. Redefining Similarities
- Shared traits reflect optimal design solutions responding to universal laws, not evidence of shared ancestry.

3. Reframing Time
- Change unfolds in phases, shaped by energy dynamics and system maturity, rather than uniformly over billions of years.

Broader Implications

1. Diversity as Design, Not Descent
- The richness of life reflects the adaptability of universal blueprints, not transformation between kinds.

2. Interdisciplinary Applications
- The same principles governing biological change apply universally—to cosmic systems, ecosystems, and even conceptual systems like thought and culture—demonstrating a **unifying blueprint** for all existence.

3. A Challenge to Traditional Assumptions
- By redefining adaptation and time, TBT invites a reevaluation of fossil records, genetic evidence, and our broader understanding of life's development and sustainability.

Conclusion: Redefining Change

The **Transcendent Blueprint Theory of Change** offers a unifying framework for understanding life's development and diversity, grounded in **universal laws** that transcend randomness:

- **Change, Not Evolution**: Life adapts purposefully within boundaries, reflecting principles of stability, balance, and efficiency.
- **Time as Dynamic and Contextual**: Change unfolds in rhythms and cycles, not on a rigid linear timescale, aligning with the energy and context of the system it governs.
- **Universal Principles**: Life's patterns are governed by laws that ensure balance, stability, and transformation across all systems—whether biological, cosmic, or conceptual.

This redefinition shifts our perspective from randomness to order, from gradual chaos to **dynamic intention**. Diversity becomes not the result of blind evolution but the expression of an **ordered, transcendent design**—a blueprint woven into the very fabric of existence, guiding life through **cycles of growth, adaptation, and renewal**.

Chapter LX: The Law of Work

Principle:
No outcome, creation, or transformation can occur without the expenditure of energy, effort, or intentional action. Work is the universal currency of progress and change.

Foundations of the Law of Work
Work is not a burden but the mechanism through which the universe moves, grows, and evolves. From the grand scales of galaxies to the smallest seed pushing through soil, progress depends on **intentional energy transfer**—moving systems from potential to reality.

The Bible states plainly:

"If a man will not work, he shall not eat." — *2 Thessalonians 3:10*

This scripture reveals a universal truth: work is essential for survival, sustenance, and growth. To refuse work is to reject participation in the dynamic processes that sustain all life and systems.

Even the origins of human labor reflect this truth. After the Fall, God told Adam:

"By the sweat of your brow you will eat your food until you return to the ground…" — *Genesis 3:19*

Work became both a necessity and a curse, a constant reminder that effort is required to produce outcomes. Yet, within this curse lies a deeper truth: work is not futile—it is the process that shapes us, refines us, and brings forth life.

Core Elements of the Law of Work

1. Energy Transfer as the Foundation

- Work is the movement of energy—physical, mental, or systemic—transforming potential into reality.
- Without work, systems stagnate, decay, and entropy overtakes order.
- *Example:* A field untended will produce thorns, not crops. The absence of effort ensures decline.

2. Effort Creates Motion

- Motion is the precursor to transformation, and work drives motion.
- Work applied to any system—be it physical, biological, or mental—initiates the processes of growth and progress.
- *Scriptural Echo: "In all toil there is profit, but mere talk tends only to poverty."* — Proverbs 14:23

3. Work as a Universal Constant

The Law of Work applies across all systems:

- **Physical**: Energy moves through thermodynamics, fueling systems from machinery to celestial bodies.
- **Biological**: Organisms expend energy to grow, reproduce, and survive.
- **Mental and Social**: Innovations, relationships, and progress arise from the effort invested in thought and collaboration.

4. Work as an Investment

- The effort we invest compounds over time, yielding exponential returns.
- Purposeful work produces order, abundance, and innovation. Conversely, misdirected or neglected effort leads to waste.
- *Scriptural Reflection: "Whoever sows sparingly will also reap sparingly, and whoever sows generously will also reap generously."* — 2 Corinthians 9:6

The Law of Work and Friction

Work rarely flows without resistance. The **Law of Friction** shows that resistance is necessary for refinement, adaptation, and strength.

- Resistance in exercise builds strength and endurance.
- Friction in challenges sparks creativity and growth.

In this way, friction enhances the outcomes of work, transforming obstacles into tools for optimization.

"The testing of your faith produces perseverance." — James 1:3

Applications of the Law of Work

1. In Nature

- Plants absorb sunlight, expend energy to grow, and convert resources into new life.
- Predators hunt to survive, expending energy for the reward of sustenance.
- Decomposers work tirelessly to recycle nutrients, ensuring the balance of ecosystems.

2. In Human Systems

- **Personal Growth**: Progress in health, relationships, or learning requires consistent effort. Neglect guarantees decline.
- **Innovation and Business**: Organizations thrive when energy (work) is directed toward high-impact actions. Misaligned effort drains resources and erodes value.

3. In Thought and Creation

- Ideas require mental work to become tangible realities. Innovation and progress are born from intentional thought and action.

4. In Spirituality

- Faith itself requires work:
- *"Work out your own salvation with fear and trembling."* — Philippians 2:12
- The principles of labor, growth, and renewal mirror the divine process of transformation.

The Curse and the Blessing of Work

While work began as part of the curse in Eden, its deeper purpose reveals itself through the *Transcendent Blueprint Theory*. Work is not punishment—it is the **mechanism of creation and refinement which was made that way at the moment of the curse.**

1. **The Sweat of the Brow: Effort connects us to the cycles of life, forcing us to participate in growth and renewal.**
2. **The Path to Redemption**: Work refines the spirit, aligns intentions, and transforms potential into purpose.

In this light, work becomes a blessing:

"Whatever you do, work at it with all your heart, as working for the Lord, not for men." — *Colossians 3:23*

Key Insights from the Law of Work

1. **Work is Inevitable**: Nothing progresses without effort. Even inaction leads to decline.
2. **Friction Enhances Work**: Resistance refines outcomes, driving growth and transformation.
3. **Purposeful Work Creates Legacy**: Aligned, intentional effort produces meaningful and lasting impact.

Integration into the Transcendent Blueprint Theory

The **Law of Work** is foundational to TBT, connecting with other universal principles:

- **Seed Law**: Potential remains dormant until activated by work.
- **Law of Magnetism**: Work generates motion, attracting aligned outcomes.
- **Law of Frictional Refinement**: Resistance refines and enhances the impact of effort.
- **Cycles and Reciprocity**: Work maintains balance and sustains the flow of energy in all systems.

Closing Thought: The Purpose of Work

The Law of Work reveals that energy, effort, and intentional action are woven into the very fabric of existence. To refuse work is to refuse growth, progress, and purpose.

Yet, when we embrace work—not as a burden but as a blessing—we participate in the **universal rhythm** that transforms potential into reality. We align ourselves with the blueprint that sustains all life, knowing that effort invested today lays the foundation for tomorrow's renewal.

"Commit to the Lord whatever you do, and your plans will succeed." — *Proverbs 16:3*

In work lies the key to transformation. To master work is to master the **mechanism of creation itself**—unlocking infinite possibilities in alignment with the blueprint of the universe.

Chapter LXX: The Law of Friction

At first glance, friction appears as a force of resistance—an obstacle that opposes motion, progress, and ease. It slows, grinds, and challenges. And yet, this resistance is not arbitrary. Friction, in its many forms, **refines systems**, reveals purpose, and drives transformation. It is both the constraint and the catalyst, the force that shapes rough edges into precision, motion into progress, and struggle into strength.

The Bible often frames resistance as a tool of refinement:

"Consider it pure joy, my brothers and sisters, whenever you face trials of many kinds, because you know that the testing of your faith produces perseverance." — James 1:2-3

Friction is not simply endured—it is harnessed. The trials we face, the challenges we resist, and the struggles we push through become the very tools that sharpen us and our systems. The Law of Friction reveals that opposition is necessary, not incidental. Without it, growth would be impossible, systems would remain unrefined, and potential would remain dormant.

The Inescapable Nature of Friction

Resistance is inevitable. It exists in every system and interaction, from the physical to the spiritual. In **physics**, friction opposes motion. A ball rolling on the ground eventually slows, its energy dissipating into heat. In **life**, ideas encounter resistance when they challenge established norms, and individuals face obstacles when striving for growth.

Resistance, though unwanted at times, is part of the **blueprint of motion**. Without it, there would be no traction, no progress, and no refinement.

Consider the sculptor with a block of marble:

- It is through friction—the grinding and chiseling—that raw stone becomes art.
- The resistance between tool and surface does not destroy; it **reveals the masterpiece hidden within**.

Friction is the sculptor of systems, shaping them into their optimal forms.

Friction as a Refiner

In a world of resistance, systems are forced to adapt or stagnate. Friction reveals flaws, inefficiencies, and limitations, compelling refinement and precision. This process of grinding away imperfection transforms the rough into the smooth, the incomplete into the whole.

- A forest fire, destructive as it seems, clears deadwood and releases seeds dormant for years. Without the fire, renewal cannot occur.
- In evolutionary biology, resistance from the environment forces species to adapt, refine, and grow stronger.
- In human lives, struggles—whether personal, professional, or spiritual—build character and resilience.

"But He knows the way that I take; when He has tested me, I will come forth as gold." — *Job 23:10*

Gold is refined by fire, sandpaper smooths wood, and friction shapes systems into their highest potential. The process is never easy, but the result is transformation.

Friction's Dual Nature: Obstacle or Tool?

The Law of Friction reveals that resistance is neither good nor bad—it is **neutral**. It becomes a tool or an obstacle depending on how we respond to it.

Too much friction can halt progress, creating chaos, dysfunction, and waste. A team burdened by conflict cannot move forward, and a machine with too much resistance overheats and breaks down.

Too little friction, however, can lead to stagnation. Systems without challenge remain unrefined and weak, untested by the forces that could sharpen them.

Balanced friction, on the other hand, drives progress. In nature, in systems, and in human lives, this balance is key.

- A river carves through rock over centuries, refining its path.
- Athletes encounter resistance in training to build strength and skill.
- New ideas meet opposition that forces refinement, strengthening their foundation and value.

Key Insight: Friction, when harnessed, becomes the sharpening stone of progress.

The Energy of Friction: Resistance as a Source of Power

While friction opposes motion, it also generates energy. In physical systems, friction produces heat, converting movement into a new form of energy. In social systems, tension—protests, debates, and conflicts—becomes the fuel for systemic change.

- The resistance of a braking system, though it slows motion, produces energy that can be harnessed.
- Social revolutions arise from the friction of dissatisfaction, creating momentum that reshapes societies.
- Personal struggles generate the energy of perseverance, transforming individuals into stronger, wiser versions of themselves.

Friction does not merely resist; it **unlocks latent energy**, channeling it toward transformation when properly understood and managed.

Applications of the Law of Friction

Friction appears everywhere, shaping outcomes across all systems:

1. **In Nature: Environmental resistance drives adaptation, clearing the way for renewal and growth.**
   - A seed must break through the soil to sprout.
   - Animals evolve in response to environmental challenges.

2. **In Personal Growth: Life's struggles are friction that builds character and strength.**
   - Overcoming adversity creates resilience.
   - Challenges reveal flaws and push us to refine skills, thoughts, and actions.

3. **In Innovation: New ideas meet resistance, forcing adaptation and improvement.**
   - Technology advances by solving friction points in existing systems.

- Creative solutions arise from the tension of problems and constraints.

4. **In Relationships and Societies: Conflict refines understanding, connection, and systems.**
   - Disagreements, when balanced, build stronger bonds.
   - Social friction—protests, reform movements—drives progress and systemic change.

Integration into the Transcendent Blueprint Theory

The **Law of Friction** is inseparable from the **Transcendent Blueprint Theory**, aligning seamlessly with other universal principles:

- **The Law of Work**: Friction enhances effort, sharpening the outcomes of work through resistance.
- **The Seed Law**: Potential cannot unfold without resistance to shape and refine it.
- **The Law of Magnetism**: Friction generates push-pull dynamics, creating motion through attraction or repulsion.
- **The Law of Neutral Freedom**: Friction balances freedom with responsibility, preventing stagnation and ensuring sustainable evolution.

Friction is not an isolated force but part of a larger, interconnected system that governs progress, refinement, and renewal.

Closing Thought: Embracing Friction

The Law of Friction reveals that resistance is not something to be avoided—it is something to be **understood, embraced, and harnessed**. Every challenge, every struggle, and every obstacle is an opportunity for refinement and transformation.

"We also glory in our sufferings, because we know that suffering produces perseverance; perseverance, character; and character, hope." — *Romans 5:3-4*

Friction does not stop progress; it sharpens it. It exposes what is weak, grinds away what is unnecessary, and polishes what remains into something strong, refined, and true.

To resist friction is to resist growth. To embrace it is to step into the universal rhythm of refinement—a process that transforms systems, ideas, and individuals into their highest potential.

Friction is the tool that turns motion into progress, resistance into strength, and struggle into transformation. In its sharpening force lies the promise of renewal, balance, and growth.

Chapter LXXX: The Flow of Alignment: Spirit, Energy, and the Blueprint

For a moment let me step away from the mundane babbling of Laws, Theories, and Hypotheses to something practical, tangible, yet misunderstood. Those moments in life where something outside of ourselves moves within us—unseen yet undeniably physical. It is energy which flows through the body, a current of electricity, both vibrant and purposeful. I first experienced this in a moment of profound faith—laying hands on someone in need of healing—and felt electricity move not through my entire body, but in a **directional flow**, starting at the top of my head and extending through my arm.

This was not imagination. It was real. And I could feel it.

That moment marked the beginning of my curiosity about what this sensation meant—spiritually, physically, and universally. Was this the filling of the Spirit itself? Or was this energy an expression, evidence, or effect of something deeper?

Over the years, it has returned in unexpected ways. Once, while speaking to a client about **trees**, of all things—not faith, not healing—I felt that same energy coursing through me. Another time, while sitting next to my wife, she could feel the energy flowing through my leg as we spent time in prayer. The electricity radiated outward. And then, there are days like today, where the fullness of the Spirit energized my **entire being**, every fiber alive, humming with what I can only describe as a electrical current.

It has become clear to me that this energy, this flow, cannot be confined to a single experience. It transcends settings, topics, or intentions. Whether in a moment of sacred prayer or a conversation about creation itself—trees, roots, soil—there is something universal at play.

Electricity and the Flow of Alignment

Energy, at its core, is movement. In the physical world, **electricity** is the flow of charged particles along a pathway of least resistance. The sensation I felt is not unlike this process:

- **The Spirit as Source**: When we are filled with the Spirit, our resistance—mental, emotional, physical—falls away. Energy flows because alignment has been achieved.

- **Directionality**: Just as electricity moves purposefully along a wire, this energy flows where it is needed—through my arm for healing, through my words when speaking, or through my body when fully present with the Spirit.

Theories and Laws in Action

When I began articulating the *Transcendent Blueprint Theory* **(TBT)** and its accompanying laws, I did not yet fully connect these principles to my own experience. Now, I see how they provide a deeper framework for understanding these moments of energetic flow.

1. **The Law of Magnetism**
   - Energy flows toward **alignment**.
   - When I laid hands on someone for healing, there was a **polarization**—a purposeful attraction of energy toward that act. The Spirit, the source of life and healing, moved through me because the moment required it.

2. **The Law of Work**
   - No transformation occurs without energy transfer.
   - In moments of healing, teaching, or even speaking about trees, I became a **conduit**. The effort, intention, and purpose were the work that allowed energy to move and create outcomes—whether physical healing, emotional connection, or clarity of insight.

3. **The Law of Neutral Freedom**
   - Balance and flow occur when resistance is removed.
   - Whether through surrender in faith, passion in purpose, or clarity of thought, I found myself in a state of **neutral freedom**—a space where energy could flow unimpeded.

Energy and Creation: Trees as a Reflection of Spirit

I cannot ignore the moment when I felt this energy while speaking about trees. Trees, after all, are not just natural organisms—they are physical manifestations living under the canopy and by the design of the universal blueprint.

- **Roots and Shoots**: Trees conduct energy, drawing it upward through their roots and dispersing it through their leaves. This parallels how energy seems to move directionally within me—drawn from a higher source and directed toward a purpose.

- **Balance and Flow**: A tree's survival depends on harmony—energy flowing efficiently through its systems. When I spoke about trees, my passion aligned me with the blueprint, allowing energy to flow through me with the same clarity and purpose.
- In that moment, it was as though the Spirit was reminding me of that same moment I had as a child in the snow, when the physical and unseen worlds blended together becoming one.

Alignment: Spirit and the Universal Blueprint

The energy I experience when filled with the Spirit, or when deeply aligned with purpose, does not originate from me. And it is not the Spirit itself, but rather a universal flow of energy allowed and directed by the Spirit. This energy exists as a principle **embedded** in creation, governed by the same laws that sustain the universe.

The *Transcendent Blueprint Theory* (TBT) reveals that alignment, through the spirit, opens pathways for this energy to flow—through systems, through purpose, and through individuals. The Spirit, being the infinite Source of all life, activates this flow with clarity, direction, and power.

Energy as Universal Law, Activated by the Spirit

Energy—like magnetism, gravity, and motion—is a universal law underwritten into existence:

- It is neutral in its potential, present in all systems.
- It becomes purposeful when directed and aligned.

Just as magnetism exists universally but manifests most clearly in a magnet, this energy flows best through aligned systems—systems where resistance is removed, and pathways are clear. The Spirit is the activating force that allows this energy to flow fully, powerfully, and intentionally.

The Spirit's Role in Alignment

Scripture describes the **Spirit** as a dynamic, life-giving force:

"Whoever believes in me… rivers of living water will flow from within them." — John 7:38

This "living water" is not electricity or energy itself but the effect of the Spirit's presence—unblocking, aligning, and activating pathways for the flow of life.

- **The Spirit as Source**: The energy does not come from me. It comes through me when I am aligned with the Spirit's purpose.
- **The Energy as Evidence**: The electricity I feel—the hum, the flow, the directionality—is not the Spirit but evidence of it and alignment. It reflects a universal law being activated under the Spirit's guidance.

The Overlap of the Spiritual and Physical

The boundary between the spiritual and physical is far thinner than we often recognize. The Spirit, as the architect of the universe, works within the very blueprint He created:

1. **In Nature: Trees, ecosystems, and stars demonstrate energy flows that align with balance and purpose.**
   - The energy I feel while speaking about trees is not separate from this truth—it is alignment with creation itself.
2. **In Human Life: Passion, purpose, and faith align us with the same energetic flow embedded into all existence.**
   - Even when I discuss trees—an act that seems secular—the Spirit reveals alignment through clarity, resonance, and flow.
3. **In Faith: The Spirit activates and magnifies what already exists, removing resistance and directing energy with divine precision.**

The energy itself—this flow—is a principle woven into the blueprint. The Spirit, however, is the conscious activator of that flow, amplifying its power and aligning it with purpose.

The Personal Revelation: When Alignment Opens the Pathways

Reflecting on the times I have felt this energy—whether laying hands for healing, speaking passionately about trees, or worshiping—I see one unifying truth: alignment opens the pathways.

- My resistance falls away.

- I become a vessel through which energy flows.
- The Spirit allows, directs, and magnifies this flow, aligning it with divine purpose.

This experience is not limited to "spiritual" moments because alignment itself is a universal law. The Spirit works through it, showing that even conversations about creation—trees, soil, roots—can reflect deeper truths and allow energy to move.

Conclusion: The Spirit Activates the Blueprint

The energy I feel—whether directional, subtle, or overwhelming—is evidence of universal laws. Those that exists underwritten into creation. The Spirit, as the Source, activates, directs, and magnifies this energy, aligning it with truth and purpose:

- **The Spirit is the Source**: Energy flows because the Spirit opens pathways.
- **The Energy is the Evidence**: A reflection of alignment and motion within the universal blueprint.
- **Alignment is the Pathway**: Through faith, purpose, and surrender, we remove resistance and allow the flow to occur.

This revelation transforms my understanding of these experiences. When I am aligned—whether in faith, work, or passion—the Spirit uses the blueprint of creation to reveal itself. The energy flows because the Spirit allows it to flow.

In these moments, I see the overlap of the spiritual and the physical, the seen and the unseen. I am not just speaking, praying, or healing—I am participating in the rhythm of creation itself.

And perhaps that is the greatest truth of all:

The Spirit fills, aligns, and flows—revealing the design embedded in all things. In these moments, I am not just observing the blueprint; I am living it. And then there is miracles which we will discuss later in this book.

Chapter LXXXI: The Law of Aligned Causality

"Everything happens for a reason." This phrase has echoed through centuries of philosophy, religion, and personal reflection, offering comfort and perspective in moments of doubt. The Bible adds an even deeper dimension: *"All things work together for good for those who love the Lord and are called according to His purpose."* While often interpreted spiritually, this statement reflects a universal truth that transcends doctrine. Through the lens of the **Transcendent Blueprint Theory (TBT)**, this concept emerges as a fundamental principle governing all systems—a **law of aligned causality**.

This chapter explores how TBT reveals the deeper meaning of this law, demonstrating that all events, regardless of scale, align with universal principles of growth, energy flow, and purpose.

A Universal Principle: Aligned Causality

The **law of aligned causality** suggests that every event, whether perceived as positive or negative, serves a purpose within the larger systems it inhabits. This law is not confined to human experiences but applies universally, shaping the dynamics of life, nature, and the cosmos.

Cycles of Cause and Growth

1. **Universal Laws at Work:**
   - Systems across all disciplines—cosmic, biological, and social—are governed by cycles of action and reaction. Every event redistributes energy, catalyzes change, and contributes to the system's overall balance and evolution.
   - In nature, a tree falling in a forest may appear destructive, but its decaying wood enriches the soil, supporting future growth.

2. **Purposeful Energy Flow:**
   - TBT reveals that energy, once set in motion, is never wasted. Whether in a galaxy or a human life, events trigger reactions that serve larger cycles of growth and renewal.

Time and Alignment

Through the **Contextual Time Hypothesis**, TBT explains how the meaning of events unfolds differently based on timing:

1. **Short-Term Perception vs. Long-Term Purpose:**
    - In the moment, events may seem random or purposeless. Over time, however, their role within the system becomes evident. For *example:*
    - A storm may initially devastate a landscape, but it also disperses seeds, clears deadwood, and creates conditions for renewal.
    - Personal struggles often lead to growth, preparing individuals for opportunities or insights they could not have achieved otherwise.
2. **Dynamic Causality:**
    - Events are not isolated; their significance shifts as systems change. What appears as chaos today may prove essential for balance and transformation tomorrow.

Evidence of Aligned Causality

Aligned causality is not a new idea, but its application through TBT provides a unifying framework for understanding it. Evidence for this law appears across disciplines:

Nature and Ecology

1. **Cycles of Renewal:**
    - In ecosystems, apparent destruction often paves the way for growth:
    - Forest fires, though destructive, activate seeds and renew the soil with nutrients.
    - Seasonal death in plants enriches the earth for the following year's growth.
    - These events are not random but serve a purpose in the larger system of life.
2. **Adaptation Through Stress:**
    - Trees subjected to strong winds develop reaction wood, making them more resilient. The stress, while challenging, strengthens the tree and aligns it with its purpose of survival and growth.

Human Experience

1. Personal Growth Through Adversity:

- Challenges often lead to profound growth, even if their purpose is not immediately clear. For instance:
- Failures can redirect individuals toward paths better suited to their talents or goals.
- Loss or hardship can foster empathy and resilience, aligning individuals with roles they might not otherwise fulfill.

2. The Role of Intention:

- TBT suggests that human thought and action align with universal principles, creating outcomes that serve larger cycles. Intentional actions ripple outward, influencing events in ways that are often unseen but purposeful.

The Cosmos

1. Galactic Systems:

- On a cosmic scale, even seemingly chaotic events like supernovae play roles in creating new stars and dispersing the elements necessary for life. These events, though violent, align with the universe's blueprint for creation and transformation. What appears random or destructive ultimately serves balance, renewal, and purpose.

The Unification of Truths: From Fragmentation to Wholeness

The **Law of Aligned Causality** is more than an observation of patterns—it is the realization that humanity's scattered understanding of causality is, in fact, part of a singular, universal design. Across time, disciplines, and cultures, fragments of this law have appeared under different names:

1. In Science:

- Physics frames causality as energy moving through systems, triggering reactions.
- The concept of **entropy** shows that systems naturally seek equilibrium, transforming chaos into balance.

- Evolutionary biology reveals that stressors refine species over time, aligning them with their environment for survival.

2. In Eastern Philosophies:

- Karma teaches that actions produce consequences, aligning individuals with cycles of growth and correction.
- Zen emphasizes that stillness and motion exist together, much like causality requires both neutrality and action to produce aligned outcomes.

3. In Faith:

- The Bible highlights purposeful causality, as seen in **Romans 8:28**: "All things work together for good for those who love the Lord and are called according to His purpose."
- Scripture reframes trials as tools of refinement and realignment, showing that resistance serves transformation:

"We also glory in our sufferings, because we know that suffering produces perseverance; perseverance, character; and character, hope." — *Romans 5:3-4*

TBT reveals that these expressions—whether scientific, philosophical, or spiritual—are not contradictions but **reflections of the same universal truth**. What science observes as natural law, and what faith describes as divine purpose, are two perspectives of the same blueprint woven into existence.

Where the Fragments Meet: The Role of Alignment

TBT's **groundbreaking contribution** lies in its ability to unify these fragmented understandings into a cohesive framework. Here, causality is no longer limited to one worldview or discipline—it transcends them:

1. Alignment as Intentional Participation:

- Humans are not passive recipients of causality; they are participants. The energy we direct, the intentions we set, and the actions we take ripple through systems, shaping outcomes in ways we may not always perceive.
- This aligns with Jesus' teaching of stewardship and intentional action: *"To whom much is given, much is required."* — *Luke 12:48*

2. **Neutrality and Balance:**
   - The universe does not favor or punish; it aligns systems toward balance and growth. This mirrors the neutrality of natural law—gravity does not discriminate, nor does karma or cause-and-effect.
   - Faith offers clarity here: those who align with divine purpose perceive meaning and fulfillment more clearly because their actions resonate with the blueprint's rhythm.
3. **The Infinite Layers of Causality:**
   - Every action exists within a web of causality, its effects rippling outward in layers we cannot fully see. What appears random or isolated is part of a greater sequence, guided by the blueprint.
   - A fallen tree enriches the soil; a personal failure opens a new path; a dying star births the elements of life itself. Seen this way, causality reflects **both physical law and divine orchestration**.

A Clarifying Perspective: Truth Beyond Exclusivity

The spiritual truth of **Romans 8:28** speaks specifically to believers, offering assurance that their struggles and challenges are not meaningless but aligned with God's purpose. However, the universal nature of TBT reveals that causality is not exclusive:

- **The principles operate for all systems**—whether seen or unseen, believed or ignored. The cycles of destruction and renewal occur in nature, in the cosmos, and in human lives because they are embedded in creation.
- What faith provides is **clarity** and **intentional alignment**—the ability to trust the process and engage with it fully, recognizing that good emerges even from adversity.

This is not exclusion but an **invitation** to participate in the design of causality with awareness. Universal laws are available to all; faith opens the door to deeper understanding.

Conclusion: A Universal Law That Transcends Boundaries

The **Law of Aligned Causality** reveals that everything—seen or unseen, chaotic or ordered—is part of a greater rhythm of growth, balance, and renewal. TBT stands as a unifying framework:

- It shows that **science** reveals the patterns of causality.
- It demonstrates that **faith** interprets causality as purpose.
- It acknowledges that **philosophy** and spirituality capture glimpses of these universal laws, even if incomplete.

Where others have seen fragments, TBT reveals wholeness—a single blueprint through which causality aligns all systems toward transformation and balance.

Implications of the Law of Aligned Causality

The **Law of Aligned Causality** reframes how we see the events of life, nature, and the cosmos:

1. **Purpose Embedded in Design:**
   - TBT reveals that purpose is not an afterthought but an inherent quality of all systems. Every action—whether seen in nature, human experience, or the stars—aligns with principles of energy flow, balance, and renewal.

2. **Randomness Reframed:**
   - Events that seem chaotic or purposeless are better understood as part of a larger design. In time, challenges often reveal their role in cycles of growth and transformation.

3. **Faith as Alignment:**
   - Faith, far from being separate from universal truths, enhances our ability to perceive and trust this process. Believers experience a heightened clarity, aligning their lives with divine purpose and universal principles.

The Broader Context: Faith, Universal Truths, and Inclusion

Romans 8:28 states:

> "And we know that in all things God works for the good of those who love Him, who have been called according to His purpose."

This verse highlights the experience of alignment:

- For believers, faith acts as a **catalyst** for understanding challenges as purposeful and transformative. It reveals the good embedded within even the most difficult circumstances.
- Yet TBT shows that this universal law—causality aligning all systems with growth and renewal—applies to all creation, regardless of belief.

The verse does not exclude others from experiencing good but emphasizes the **depth of alignment** and fulfillment that comes from actively participating in the blueprint's rhythm. In nature, systems thrive in cycles of destruction and renewal, even without intention. For humans, alignment— whether conscious or not—brings purpose into focus.

Conclusion: Discovering Purpose in the Blueprint

The **Law of Aligned Causality** reveals a profound truth: nothing is random. Every event, every moment of resistance, and every ripple of energy aligns with a greater design—a design that reflects balance, renewal, and purpose.

Through the *Transcendent Blueprint Theory*, we see that science, philosophy, and faith are not opposing forces but complementary lenses. Together, they point to the same universal rhythm: causality works for good, aligning systems toward growth, balance, and transformation.

For those with faith, this truth offers clarity, trust, and hope. For all creation, it is evidence of a design that transcends chaos and binds everything—seen and unseen—into a single, purposeful whole.

This is the promise of aligned causality:

- What feels broken is being refined.
- What seems chaotic is leading to balance.
- What appears as loss prepares the way for renewal.

The blueprint is not outside of us; it flows through us. Every challenge, every choice, and every moment aligns with a purpose greater than we can often perceive.

"Everything happens for a reason" is not a simple phrase—it is a reflection of the eternal truth that energy, causality, and purpose guide all existence, inviting us to align with the universal design.

Final Transition

Thus concludes the X Chapters: a reframing and a foundation for the understanding of systems, cycles, and truths that will resonate for a thousand years to come. Now we move closer to the heart of it all: **Faith, Us, and Everything Else**—the part of the blueprint where we discover our role in its unfolding rhythm.

Part 3
Interpretations

Chapter 11: Faith — Transcendence Beyond Belief

Introduction: Exploring the Unseen
This chapter ventures into the realms of faith, miracles, and the extraordinary—not as definitive proofs of the *Transcendent Blueprint Theory* (TBT), but as contemplations and interpretations that invite further exploration. These phenomena, often described as supernatural or beyond comprehension, challenge us to consider whether they are truly outside the scope of universal laws or simply manifestations of principles we do not yet fully understand.

A Necessary Disclaimer:
The ideas presented here are not assertions of absolute truth or scientific certainty. Instead, they are possibilities—interpretive frameworks that align with the TBT's principles of cycles, reciprocity, transformation, motion, and connection. By examining these phenomena through the lens of the TBT, we aim to inspire curiosity and deeper inquiry, rather than claim definitive evidence.

While the discussions that follow are speculative, they are grounded in the idea that the extraordinary might be embedded within the ordinary, waiting to be understood. As such, this chapter serves as an invitation to imagine the broader implications of the TBT and to explore the potential that lies within the unknown.

Additional Disclaimer: A Foundation for Exploration
Before we proceed, I must also offer a word of caution. This chapter will explore concepts of faith, not to challenge or diminish ones faith, but to examine their alignment with the *Transcendent Blueprint Theory* (TBT). This theory, while encompassing and transcending individual belief systems, does not aim to overthrow the concept of faith or divinity. Instead, it seeks to illuminate how universal principles—cycles, reciprocity, transformation, motion, and connection—may underlie the miraculous and the spiritual.

Some readers may find this perspective challenging, as it proposes that healing, miracles, and faith may be manifestations of universal laws yet to be fully understood. However, this exploration does not contradict the existence of God or the spiritual realm. On the contrary, it suggests that these phenomena may be woven into the very fabric of existence, written into the "DNA" of life itself.

What if these extraordinary events are not violations of natural laws but expressions of order—a glimpse of universal truths we have yet to comprehend fully? With this in mind, I invite you to embrace this exploration, not as a critique of faith but as an opportunity to uncover the profound threads that connect us all.

Free thinking at its finest.

Let's Start With The Seed: Faith and Potential

Faith begins as a seed—small, latent, and containing immense potential. Unlike physical seeds that grow within soil, faith's growth operates across intangible and tangible domains, governed by universal laws. Its transformative power lies in its ability to expand beyond its origin, creating ripples that transcend disciplines.

1. Latency and Germination: A Spark of Potential

Faith, like a seed, exists initially as a spark—a belief, word or intention waiting for the right conditions to germinate.

- **Unique Angle**: This stage is often overlooked because it appears invisible. However, the TBT highlights that latent potential is part of all systems.
- **Example:** The faith of a parent instilling hope in a child or a moment of prayer before an unexpected resolution illustrates how unseen forces (belief or trust) initiate action and outcomes.

2. Faith as a Dynamic Root System

Once germinated, faith mirrors a root system—it seeks nourishment, draws connections, and grounds individuals or communities. Its growth is unseen yet vital to its survival. Yes, faith must grow.

- **New Perspective**: Roots are not static; they expand, adapt, and form symbiotic relationships. Faith behaves the same way, thriving through trust, relationships, and shared experiences.
- **Example:** A person's faith strengthens over time as life events— both positive and challenging—nourish it, much like a tree's roots deepen with every passing season.

3. Universal Law in Action: The Seed's Expansion

Faith aligns with cycles of reinforcement and growth, yet its progression transcends disciplines. A seed is not just a plant's beginning but a system that multiplies its impact over time.

Expanded Insight: Faith, though deeply personal, holds the potential to spark movements, healing, and transformation on a larger scale. It aligns with the TBT principle of motion, where small beginnings ripple outward, creating larger impacts. In Christian theology, the concept of "predestination" hints at the inherent potential of a seed—what it is destined to become, shaped by unseen "water." This unseen water may be different things to different people, and also grow different things, but it reinforces TBT's transcendence across disciplines and the reality that all things are subject to universal laws.

- **Example:** The parable of the mustard seed in the Bible, where the smallest seed becomes a tree that shelters others, reflects the transformative power of faith.

Key Insight: Faith as a Transcendent System

By framing **faith as a seed**, we reveal its alignment with the **TBT's universal laws**. Faith, like a seed, carries **encoded potential** that requires planting, nurturing, and time to unfold. It is not static but operates as a dynamic force capable of **transformation, motion, and expansion** beyond its original form.

Paul beautifully captures this universal truth in **1 Corinthians 3:6**:

"I planted the seed, Apollos watered it, but God has been making it grow."

This reflects the **interconnected roles** within the process of transformation:

1. **One person plants—initiating the act of faith, truth, or action.**
2. **Another waters—nurturing and supporting growth.**
3. Yet it is **God who gives the increase**—activating and fulfilling the potential encoded within the seed.

This mirrors the TBT principles of **reciprocity** and **cycles**: growth requires motion, energy flow, and a cooperative system where each role contributes to the outcome.

- **Faith as Potential**: Faith begins as a seed—small, but containing infinite possibilities.
- **Faith in Action**: Growth occurs when faith is acted upon, nurtured, and aligned with purpose.
- **The Increase**: The ultimate transformation—**the harvest**—is not entirely within human control but is amplified by a greater system of alignment, balance, and divine design.

The TBT Connection

The **Law of Work** and the **Seed Law** further clarify this process:

- **Work** initiates the planting and watering—effort directed toward transformation.
- The **Seed Law** ensures that growth is embedded in the seed's design, waiting to be activated under the right conditions.

Paul's words remind us that while humans participate in the process, the growth itself—**the increase**—reflects a universal law at work, transcending individual efforts and revealing the interconnected rhythm of creation.

The Healing Benefits of Faith: Aligning with Universal Rhythms

Faith has long been associated with profound healing effects on the mind, body, and spirit. Viewed through the lens of the *Transcendent Blueprint Theory* (TBT), faith emerges as a force that aligns individuals with universal principles, enhancing cycles of renewal, reciprocity, and transformation. Far from being a passive belief, faith actively engages the human system, creating conditions for healing and balance.

Faith and Physical Healing: The Mind-Body Connection

Modern science has increasingly validated the role of faith in physical health. While the mechanisms remain partly unknown, the connection between belief and healing is undeniable.

- **The Placebo Effect:** Studies show that patients who believe in the efficacy of a treatment, even when it's a placebo, often experience measurable improvements in health. This highlights the transformative power of belief itself, demonstrating how intention and focus can direct the body's healing processes.

- **Stress Reduction:** Faith practices, such as prayer, meditation, and religious services, lower stress levels by calming the mind and reducing cortisol production. This enhances immune function and overall resilience.

- **Cycle of Renewal:** Faith itself fosters a sense of hope, which triggers renewal even in the face of adversity. Hope reduces despair, creating a positive feedback loop that promotes physical recovery.

Example: Cancer patients with strong faith or spiritual support networks often report higher quality of life and improved outcomes compared to those without such support, suggesting that faith may influence healing pathways.

Faith and Mental Health: A Source of Stability

Faith offers profound benefits for mental health, providing stability, purpose, and a sense of connection. These effects align with the TBT principles of balance and motion, showing how faith sustains emotional and psychological well-being.

- **Reduction of Anxiety and Depression:** Faith instills hope and reduces uncertainty, providing an anchor during difficult times. Believers often feel that their challenges have purpose, alleviating existential stress.

- **Community Support:** Faith communities create networks of care and connection, fostering emotional stability through reciprocity and shared intention.

- **Resilience in Adversity:** Faith encourages a shift in perspective, framing struggles as part of something much larger that has imprinted in it cycles of growth and renewal. This enhances resilience, reducing the psychological impact of setbacks.

Example: Studies on veterans with PTSD show that those who engage in spiritual journeys, or lean on faith, report fewer symptoms and stronger coping mechanisms compared to those who do not.

Faith and Emotional Healing: Releasing Grief and Trauma

The act of faith often includes rituals, prayers, and practices that promote emotional release and healing. These practices mirror the cycles of life, allowing individuals to let go of pain and embrace renewal.

- **Forgiveness as a Healing Act:** Faith often emphasizes forgiveness, which frees individuals from the emotional burden of anger or resentment. This emotional transformation restores balance and fosters inner peace.
- **Healing Trauma:** Faith provides a framework for processing trauma, helping individuals find meaning and purpose in their experiences. This sense of meaning acts as a salve, reducing emotional scars and promoting closure.

Example: People who attend faith-based grief counseling often report faster emotional recovery and a greater sense of peace compared to those who process grief in isolation. (The same can be said to be true about addiction recovery)

Faith and Interpersonal Healing: Restoring Connection

Faith bridges the gap between individuals, fostering reconciliation and strengthening relationships. The principle of reciprocity is evident in how faith encourages giving and receiving, mending broken connections.

- **Building Trust:** Faith traditions often teach values like honesty, compassion, and service, which restore trust and harmony in relationships.
- **Creating Community:** Shared faith creates a collective identity, allowing individuals to overcome divisions and unite around common values.

Example: Faith-based initiatives in post-conflict regions often play a key role in reconciliation, helping communities heal from deep divisions through shared rituals and dialogue.

Faith as a Catalyst for Spiritual Healing

At its core, faith addresses the human need for purpose and connection to the originator. This transcendent connection heals the spirit, fostering a sense of wholeness and alignment.

- **The Principle of Transformation:** Faith transforms despair into hope, fear into courage, and isolation into belonging. These spiritual shifts mirror the universal principle of renewal, showing how faith can guide individuals through even the darkest times.
- **Miracles of the Spirit:** Many people report spiritual experiences that transform their lives, often attributed to moments of deep faith. These experiences, though intangible, provide profound healing by aligning individuals with the underlying universal blueprint.

Example: Near-death experiences often lead to transformative faith journeys, with individuals reporting a newfound sense of peace, purpose, and connection to the world.

Faith and the Universal Blueprint

Faith, through the lens of the *Transcendent Blueprint Theory* (TBT), emerges as a profound force that transcends disciplines while remaining deeply rooted in universal laws. It is not merely abstract or intangible; rather, it is a dynamic force that follows the same principles of cycles, reciprocity, transformation, motion, and connection that govern all of existence.

Faith reflects the rhythms of life and creation, operating as a bridge between the seen and the unseen. It transforms belief into action, intention into healing, and hope into renewal. Like thoughts and words, faith interacts with tangible matter, offering a way to influence and align with the greater systems of life.

Faith, when placed correctly, acts as a force of renewal and connection. Its expressions may vary, but its impact consistently mirrors the patterns of TBT. Faith transforms the intangible into the tangible, shaping outcomes that guide healing, growth, and restoration.

Faith exists at the intersection of the personal and the universal. While deeply personal in its practice, faith also connects individuals to something greater. It operates as a dynamic system, offering a glimpse into how belief may interact with and amplify the natural rhythms of existence.

This chapter delves into faith as a dynamic force—one that operates within the principles of TBT to illuminate how belief, action, and transformation intersect with universal laws. By exploring its capacity for renewal and connection, we uncover the ways faith exemplifies the *Transcendent*

Blueprint Theory in action, revealing its role in both individual lives and the greater systems we inhabit.

Faith Through the Lens of TBT: Universal Principles in Action

Faith is often seen as an abstract or personal experience, yet when examined through the *Transcendent Blueprint Theory*, it emerges as a dynamic system cemented into the fabric of our existence by universal principles. Cycles, reciprocity, transformation, motion, and connection shape faith as profoundly as they shape physical, biological, and societal systems. This section explores the mechanics of faith, unveiling how it transcends disciplines and operates within the universal blueprint, offering profound insights into its role in human existence.

1. The Cycles of Faith: Patterns of Renewal

Faith mirrors the cycles observed in ecosystems and the cosmos, characterized by ebb and flow, growth, and renewal. When viewed through the lens of the *Transcendent Blueprint Theory*, faith emerges as a dynamic organism or system. Its inherent adaptability to personal and societal needs demonstrates the universal principles of cycles and resilience.

- **Personal Cycles**

Faith often begins as a spark, ignited by inspiration, crisis, or a search for God. Like an organism responding to environmental stress, it grows through reinforcement and reflection, occasionally waning before renewing in a transformed form.

Example: A person facing grief may lose faith temporarily, but through relationships and moments of reflection, they rediscover it in a deeper, more resilient way. This mirrors dormancy and regrowth in ecosystems, such as trees shedding leaves and renewing through seasonal cycles.

- **Cultural and Historical Cycles**

Faith traditions, like biological systems, adapt to cultural and societal changes to remain relevant. Periods of spiritual dormancy often precede revival, much like ecological systems recover after disruption.

Example: The "Jesus Movement" of the 1970s in the United States, led by various individuals and groups independently, was a modern spiritual revival. It renewed faith among younger generations and adapted religious practices to align with cultural shifts. This mirrors the way ecosystems regenerate and thrive after disturbances, such as forests recovering after wildfires. It also highlights humanity's innate need for spiritual renewal in response to stagnation, a pattern echoed in biological systems where disruption often catalyzes cycles of renewal and growth.

Faith thrives on cycles of creation, reinforcement, and renewal, aligning with the TBT principle that no system remains static. Its cyclical nature, whether personal, societal, or cultural, reflects humanity's innate response to disruption and stagnation, offering resilience and continuity. These patterns underscore faith's adaptability within the universal blueprint, mirroring the regenerative cycles observed in nature and affirming its role in sustaining equilibrium across systems.

2. Reciprocity in Faith: The Dynamic Exchange

- Faith operates as a reciprocal system, where trust, action, and outcome exist in a continuous exchange. This dynamic interplay mirrors the universal laws outlined by the TBT, demonstrating that reciprocity is not limited to biological or physical systems but extends to intangible realms like belief and spirituality.

Reciprocity in Belief Systems
Faith traditions emphasize the balance of giving and receiving, where devotion, holiness living, or prayer aligns with outcomes like blessings, enlightenment, or peace. These cycles reflect the same principles that sustain natural ecosystems.

Example: Acts of service or charity within a community create trust and connection, mirroring the nutrient exchange in mycorrhizal networks, where plants provide carbon to fungi in exchange for essential nutrients. This natural flow ensures mutual benefit and reinforces balance.

Faith as Energy Exchange
Faith functions as an energetic system, where shared rituals, collective intention, and communal actions amplify individual and group outcomes. These exchanges align with the physical principles of energy transfer and amplification.

Example: The Amplification of Collective Alignment

A congregation united in prayer releases **collective energy** through the Spirit that amplifies the effects of individual intentions. This mirrors the way particles in a **magnetic field** synchronize to enhance their combined strength—what one particle alone cannot achieve becomes possible through alignment.

Jesus Himself articulated this principle in **Matthew 18:19-20**: *"Again, truly I tell you that if two of you on earth agree about anything they ask for, it will be done for them by my Father in heaven. For where two or three gather in my name, there am I with them."*

This verse reflects both a **spiritual promise** and a universal truth:

- Alignment magnifies impact. Just as particles align in a magnetic field or waves synchronize to amplify force, **agreement among people amplifies energy, intention, and results**.
- The Spirit, far from bypassing natural systems, works **in harmony with universal laws**, guiding and amplifying the rhythms and flows already present in the blueprint.

This perspective does not diminish the role of the divine, but instead reveals its coherence with the *Transcendent Blueprint Theory* **(TBT)**:

- The **Law of Work**: Intentional energy (prayer, action) is directed with purpose.
- The **Law of Magnetism**: Alignment—whether spiritual, mental, or physical—creates a **force field** greater than the sum of its parts.
- The **Seed Law**: When faith and action are combined, the encoded potential within the moment begins to unfold.

The Intersection of Spiritual and Physical Dimensions

This unified view suggests that the spiritual and physical dimensions do not compete with one another; they **intersect and reinforce** each other. Prayer, faith, and collective action activate **natural laws** embedded into creation:

- When two or more gather and align their intentions, they activate principles of **resonance, reciprocity, and amplification**—laws present within the universe itself.

- Yet, the Spirit remains the **guiding force**—the activator that moves within these systems, transforming alignment into outcomes beyond human ability.

Transcendence of Universal Laws

Reciprocity in faith exemplifies how universal laws transcend disciplines, applying equally to physical, biological, and intangible systems. Faith, like all systems, follows cycles of input and output, motion and balance. The same principles that govern nutrient flows in ecosystems or the exchange of energy in physical systems apply to the give-and-take inherent in belief. This alignment underscores the universality of TBT.

Reciprocity in faith is not merely a spiritual concept but a universal principle governed by natural laws. Whether expressed through communal worship or personal acts of service, faith operates as a flowing exchange of energy, trust, and renewal, reflecting the TBT's assertion that all systems—tangible or intangible—are interconnected and guided by the same foundational blueprint.

3. Transformation Through Faith: Shaping Reality

Faith is a **catalyst for transformation**, capable of reshaping individuals, communities, and entire societies. Much like the principles of transformation and renewal found in the *Transcendent Blueprint Theory* **(TBT)**, faith initiates processes of adaptation, growth, and resilience in response to challenges and opportunities.

Personal Transformation

Faith enables individuals to transcend adversity, not merely by reframing challenges but by **activating profound change**. This aligns with the TBT principle of transformation, where systems evolve and adapt to ensure their continuity and resilience.

- **Example:** Stories of addiction recovery through faith illustrate how belief fundamentally alters individuals, reshaping their **emotional, mental, and even physiological frameworks**. This process mirrors **cellular regeneration** in biology, where damaged systems repair and renew to sustain life, reflecting the universal law of renewal.

Societal Transformation

Faith-driven movements have historically acted as forces of societal change, uniting people under shared ideals and fostering large-scale transformation.

Faith reflects the natural processes of **transformation and renewal**, acting as a catalyst for both personal and collective change consistent with the unseen blueprint. Its ability to initiate change aligns with TBT principles of **motion, connection, and reciprocity**. This demonstrates how belief systems transcend disciplines and manifest tangible outcomes in alignment with universal laws.

4. Motion and Friction in Faith: Dynamics of Belief

Faith is not static—it is a **dynamic force** that grows through motion and evolves through friction. When viewed through the lens of TBT, faith mirrors the physical principles of **energy and motion**, changing through cycles of challenge, refinement, and alignment.

The Motion of Faith

Faith propels individuals and communities into action, creating ripple effects that extend far beyond their point of origin. This mirrors how **energy waves** propagate in physics, spreading influence across systems.

- **Example:** A missionary spreading their message generates ripples of influence, much like energy waves traveling through a medium. These ripples carry the potential to **transform distant communities**. When aligned with purpose, this motion sustains momentum and fosters growth.

Friction and Refinement

Resistance—whether doubt, adversity, or criticism from others—it acts as **friction**, refining faith through challenge. Like physical friction that enhances motion by removing inefficiencies, faith is strengthened and clarified through opposition.

- **Example:** Historical debates between science and religion have not only refined both fields but also illuminated areas of **overlap and connection**. This interplay mirrors the **dynamo effect**, where motion against resistance generates **usable energy**.

Faith, like motion in physics, thrives on **energy, resistance, and refinement**. Its refinement through friction aligns with the universal rhythms of growth and transformation. These dynamics provide tangible evidence that the principles of TBT transcend disciplines, governing all systems.

5. Connection: Faith as a Universal Bridge

Faith is a binding force that connects **individuals, communities, and the unknown**. Through the TBT lens, faith aligns with the universal principle that no system exists in isolation. Its power to bridge gaps—between people, ideas, and the intangible—mirrors the connections observed in natural and physical systems.

Interpersonal Connection

Faith strengthens **social bonds**, fostering resilience and collaboration within communities. These connections mirror the **structural networks** found in ecosystems, where interdependence sustains the system's health.

- **Example**: Religious festivals, communal prayers, and shared spiritual traditions unite individuals across boundaries. This collective alignment mirrors the **mycorrhizal networks** in forests, where interconnected roots share nutrients and information, enhancing the system's resilience.

Faith as a Bridge to the Unknown

Faith in God serves as a bridge to navigate **uncertainty**, linking the tangible to the intangible and the known to the unknown. This reflects the TBT principle of **connection**, where systems transcend boundaries to interact with larger frameworks.

- **Example:** Belief in life after death or judgment coming based on truth reflect our need to connect to the broader design of existence. Through faith, individuals engage with the supernatural, aligning their beliefs with the **natural rhythms** and principles of the *Transcendent Blueprint Theory*.

This comparison does not seek to diminish or explain away ones faith. Instead, it reframes faith as a **necessary part of our existence, drawn into the universal blueprint**, revealing its harmony with the patterns that sustain all systems.

Faith serves as a **bridge to the entire existence**, reflecting TBT's principle of **interdependence**. By connecting individuals to one another and to the unknown, faith demonstrates how systems thrive through **connection and collaboration**, offering a tangible example of the universal laws that govern all existence.

Faith Beyond Belief: Miracles as Universal Possibility

Miracles often inspire awe because they seem to defy natural laws. However, through the lens of the *Transcendent Blueprint Theory* **(TBT)**, miracles may not break these laws but rather reveal dimensions of the blueprint we are yet to fully comprehend. They may represent the transcendence of universal principles across disciplines or the functioning of deeper, hidden laws woven into the cosmos.

This view does not discredit their divine or spiritual origin. Instead, it highlights their broader application within the framework of existence. Faith may play a crucial role in harnessing these principles, but miracles are not limited to any single faith tradition. Even the Bible and history reflect accounts of miracles performed by those outside a specific faith, suggesting that the potential for the miraculous is embedded in the blueprint of life itself.

- **Miracles and the Blueprint of Creation**

Miracles align with the cycles, motion, and transformation principles of TBT, demonstrating possibilities inherent within the universe. They reflect not only divine intervention but also the profound intricacy of the universal laws that govern our existence.

  - **Example:** Healing miracles resemble regenerative processes in ecosystems. Just as a forest recovers after a fire through cycles of renewal, so too may miraculous healing reflect processes yet to be fully understood within the laws of biology, physics, or thought.

- **Faith as a Catalyst for Miracles**

Faith amplifies intentionality and focus, serving as a channel through which the miraculous is often realized. However, miracles may transcend individual belief, functioning as evidence of universal principles accessible to all.

- **Example:** Biblical accounts of miracles performed by those outside the faith illustrate that miracles are not bound solely by belief but operate within a broader framework, possibly governed by laws beyond our current understanding.

Miracles, when viewed through TBT, challenge conventional perspectives. Again if the *Transcendent Blueprint Theory* is true, then it applies to all things. Miracles are not necessarily violations of universal laws but manifestations of deeper principles possibly encoded within the blueprint. This perspective does not diminish their origin but reveals a universe so intricately designed that its principles extend seamlessly into the extraordinary.

Closing Thought

Miracles are extraordinary, yet they are not detached from the systems that sustain existence. They serve as profound evidence of universal laws that transcends human comprehension and are aligned with the principles of the *Transcendent Blueprint Theory*. This understanding enhances their origin and significance, suggesting that the miraculous is a phenomenon humanity is entrusted to explore, understand, and with it align.

Chapter 12: The Carbon Question

The Role of Carbon in Earth's Climate: The Contemporary Perspective

Carbon, an essential element for life, moves through Earth's atmosphere, oceans, soil, and living organisms in a continuous exchange known as the carbon cycle. This process maintains balance within ecosystems and supports life. However, many in the scientific community and society believe that human activity over the past few centuries has influenced this balance, leading to changes in Earth's climate.

A widely held view attributes rising levels of atmospheric carbon dioxide (CO_2) to industrial activities, such as burning fossil fuels, deforestation, and large-scale agriculture. These activities are believed to release stored carbon into the atmosphere, contributing to what is called the "greenhouse effect." This natural process, where greenhouse gases trap heat and regulate Earth's temperature, is thought to intensify as CO_2 levels increase, potentially causing global warming and changes to climate patterns.

Supporters of this perspective point to data showing a steady rise in atmospheric CO_2 concentrations since the Industrial Revolution, alongside observed increases in global average temperatures. They argue that these trends correlate closely with human activity and may result in effects such as melting ice caps, rising sea levels, and more frequent extreme weather events.

Carbon sinks, such as forests, oceans, and soil, play a key role in absorbing and storing CO_2. However, many believe that deforestation, ocean acidification, and soil degradation have diminished these sinks' effectiveness, exacerbating the issue. Methane (CH_4), another greenhouse gas released through agriculture and natural processes, is also considered a significant factor due to its high heat-trapping potential, albeit over shorter timescales.

Policy measures like the Paris Agreement aimed to address these concerns by reducing emissions, promoting renewable energy, and restoring ecosystems. Advocates view these efforts as critical for mitigating climate risks, while skeptics highlight uncertainties in climate models and the potential for natural variability to play a larger role than currently assumed.

Opposing perspectives within society point out that Earth's climate has always been subject to fluctuations driven by natural forces, such as solar activity, volcanic eruptions, and ocean currents. They question the extent to which human activities influence global climate and emphasize the complexity of the carbon cycle, which includes interactions not yet fully understood.

While debates persist, there is broad agreement on the importance of understanding the carbon cycle and its role in sustaining life. Carbon tells a story of interconnected systems, revealing the delicate balance that governs Earth's processes. Through the lens of the *Transcendent Blueprint Theory* (TBT), we uncover a new framework for understanding carbon—not as a problem to be solved, but as a reflection of the truth embedded within universal laws. By starting with TBT's "truth" principle, we gain clarity in identifying causes and pathways for meaningful change.

1. Carbon Through the Lens of Biology and Physics

Biological Lens: Carbon as the Foundation of Life

- Carbon is **the building block of life**, cycling through ecosystems as part of energy transfer. Its movement is not a problem to be solved but a **dynamic, self-regulating system** that maintains balance in the biosphere.
- **Photosynthesis and Respiration:** Carbon cycles between plants and animals through these processes, reflecting reciprocity. Plants absorb carbon dioxide, converting it into energy stored as carbohydrates, while animals release it back into the atmosphere.
- **Insight:** Disruptions to these biological cycles (e.g., deforestation) shift carbon flows but do not inherently "break" the system. The Earth adapts by redistributing carbon into other reservoirs, such as oceans or soil.

Physics Lens: Carbon as Energy in Motion

- Carbon is **a carrier of energy**, its molecules transforming between states to power systems. From combustion (releasing stored energy) to sequestration (storing energy in stable forms), carbon's role in **energy transfer** aligns with the laws of motion and transformation.
- **Kinetic Energy of Carbon:** Combustion releases carbon from fossil fuels, transferring stored energy into motion (e.g.,

powering machines). This process reflects **motion and inertia**, where the release of energy propels systems forward.

- **Potential Energy of Carbon:** Carbon sequestered in plants, soil, and oceans represents potential energy, waiting for the right conditions to be released or transformed.

- **Insight:** Rather than viewing atmospheric carbon as "excess," it can be understood as energy redistributing itself within the system, much like a pendulum swinging back toward equilibrium.

Reframing Carbon

- **Carbon as Balance:** Carbon does not disrupt balance; it **is** balance, shifting between reservoirs (atmosphere, soil, ocean) in response to system changes. For *example:*

- Increased atmospheric carbon signals reduced capacity in terrestrial or oceanic sinks, much like a river overflowing when its banks are constrained.

- These fluctuations are not failures but evidence of the system's adaptability under the universal principle of **reciprocity**.

- **Carbon as a Medium for Adaptation:** The movement of carbon reflects the **laws of motion, transformation, and connection**, transcending disciplines:

- In **biology**, carbon drives photosynthesis and respiration.

- In **geology**, carbon transitions between rock formations, soil, and fossil fuels.

- In **physics**, carbon molecules embody energy transfer, demonstrating the principle that energy cannot be destroyed but only transformed.

Refinements for Carbon Understanding and potential impact to Climate Change through Transcendent Blueprint Theory

1. Carbon as the Medium of Connectivity

- **Expanded Potential:** Explore carbon as not just a "messenger" but a network facilitator, akin to the Earth's nervous system. Investigate whether isotopic or molecular variations in carbon carry information that ecosystems use to adapt collectively. This might reveal carbon's role in synchronizing system-wide responses to environmental stress.

2. **Carbon's Role in Temperature Stabilization**
   - **TBT Expansion:** Propose examining whether carbon interacts with Earth's magnetic or gravitational fields in ways that subtly regulate energy flows. Could there be an unknown feedback mechanism at play where carbon concentrations buffer planetary energy beyond heat retention?

3. **Carbon as a Latent Catalyst**
   - **Enhanced Insight:** Carbon-rich environments may harbor regenerative potential not only in microbial recovery but also in dormant genetic material in ecosystems. Investigating carbon's role as a biochemical trigger for sudden biodiversity surges could provide new insights into ecological renewal.

4. **Carbon's Dance with Time**
   - **New Angle:** Consider carbon as a medium that records time through isotopic imprints, connecting past, present, and future Earth systems. This temporal aspect could help decode long-term cycles such as glaciations, mass extinctions, or planetary heating and cooling periods.

5. **Carbon as Reciprocity in Motion**
   - **Deepened Insight:** Reframe excess atmospheric carbon as a natural recalibration. Explore whether increased emissions might actually seed new forms of carbon sinks over time, such as unrecognized processes in the deep ocean or subterranean biospheres.

6. **Carbon's Creative Matrix**
   - **Expanded Inquiry:** Highlight carbon's role in creating entirely new chemical systems. Could its adaptability inspire biomimetic technologies or novel energy systems that align more closely with natural processes? Carbon might hold the blueprint for sustainable innovation.

7. Carbon as the Teacher of Universal Balance

- **Philosophical Depth:** Use carbon's cyclical journey as an allegory for humanity's relationship with resources. Excess carbon becomes a lesson in interdependence—showing how imbalances create opportunities for system-wide learning and adaptation.

Reframing the Current Climate Change Narrative

- Instead of treating carbon as the villain in climate change, TBT reframes it as an indicator and agent of systemic adaptation. Its "problematic" fluctuations may be signals for transformation rather than crises to eliminate.

- The prevailing focus on human-driven carbon emissions risks overlooking how carbon serves as Earth's feedback mechanism, prompting adaptation and realignment with universal laws. This blindness to carbon's broader role perpetuates the perception of a broken system, rather than one recalibrating to balance.

A View of Carbon Sinks and Their Contributions:

1. Oceans as Carbon Sinks

- The oceans absorb about **25-30% of annual human CO_2 emissions**.
- Phytoplankton photosynthesis sequesters carbon, much like terrestrial plants.
- **Total oceanic carbon sink potential**: Estimated at 37,000 gigatons (Gt) stored in deep waters.
- **Current annual absorption**: ~2.4 billion tons of carbon annually.

2. Soil and Land-Based Carbon Storage

- Soil is a significant carbon sink, storing **2,500 Gt of carbon**, far exceeding the atmosphere (~800 Gt).
- Agricultural practices, deforestation, and soil degradation reduce this potential by releasing stored carbon.
- Soil can better function as a regenerative sink if replenished through proper land-use practices (e.g., no-till farming, rewilding).

3. Atmospheric Carbon Levels

- Human activities have released **~1,500 Gt of carbon** since the industrial revolution.
- The atmosphere now holds around **870 Gt of carbon**, up slightly from pre-industrial levels (~600 Gt).
- While the atmosphere is often seen as a "problematic sink," TBT encourages viewing fluctuations as part of a dynamic adaptation process.

4. Forests and Vegetation

- Forests store ~550 Gt of carbon in aboveground biomass and absorb **~20.3 billion tons annually** (in optimal conditions).
- Deforestation has led to the release of **~1.5 billion tons annually**, diminishing their sequestration capacity.

Total Carbon Dynamics and Math:

Annual Human Emissions:

- Fossil fuel emissions: **11.34 billion tons of carbon annually**.
- Deforestation emissions: **1.5 billion tons of carbon annually**.
- **Total human emissions**: ~12.84 billion tons of carbon annually.

Sequestration by Sinks:

- **Oceans**: Absorb ~2.4 billion tons annually.
- **Soil**: Absorbs ~3.5 billion tons annually.
- **Forests**: Absorb ~20.3 billion tons annually (if undisturbed, likely less due to degradation).
- **Total sequestration potential (current conditions)**: ~26.2 billion tons annually.

Insights from the Math:

1. **There Is No "Excess Carbon" in Absolute Terms:**
   - If all sinks operated at their full potential, they could offset current human emissions.

- Current "excess" atmospheric carbon reflects a distribution alignment, not an absolute surplus, aligning with TBT principles of dynamic reallocation.

2. **Fossil Fuel Emissions Alone Do Not Overwhelm the System:**
   - Emissions are significant but not necessarily catastrophic if natural sinks are preserved and enhanced.
   - Forest loss and soil degradation are primary factors exacerbating imbalances.

3. **Other Sinks Are Overlooked in Popular Narratives:**
   - Forests dominate the conversation, but oceans and soils are equally critical.
   - Current ideologies focus on forest restoration as a singular solution, ignoring the broader system.

TBT Perspective on the Carbon Problem:

- **Cycles and Redistribution**: Carbon "excess" is part of a cycle that signals shifts in Earth's adaptive system, not an apocalyptic imbalance.
- **Reciprocity and Adaptation**: While human activity disrupts carbon flows, the Earth responds dynamically through feedback mechanisms.
- **Restoration Through Alignment**: Solutions lie not in attempting to control carbon but in encouraging systems—forests, soil, oceans—to continue their natural roles as regulators.

Key Takeaway:

The current narrative surrounding forests and fossil fuel emissions often simplifies the complexities of the carbon cycle. TBT reveals that carbon is not inherently a problem but a dynamic participant in a self-regulating system. The relatively small difference between pre-industrial and current atmospheric carbon levels indicates that the system still functions effectively, even under significant human influence. Rather than framing carbon as "excess," TBT encourages a focus on realignment—restoring disrupted cycles and ensuring all sinks operate in harmony. The solution lies in understanding carbon's role within the universal blueprint, highlighting balance rather than control.

Solving the Carbon "Problem" Through the Lens of TBT

When viewed through the *Transcendent Blueprint Theory* (TBT), the carbon problem transforms from a catastrophic issue into a dynamic process of realignment. By stepping back and observing the interconnected roles of carbon across disciplines, we uncover that the system is not broken, but misunderstood. The key to understanding and addressing the carbon question lies in recognizing the natural principles at play rather than imposing artificial narratives or control.

Let's recap.

1. Reframing Carbon: Not a Problem but a Signal

Carbon fluctuations are not failures of the system; they are its way of communicating. Carbon acts as both a medium of energy flow and a signal of systemic shifts or misalignments.

- **TBT Insight:** Carbon "excess" reflects dynamic reallocation within the self-regulating Earth system. For instance:
- **Increased atmospheric carbon reflects dynamic realignment** within Earth's interconnected carbon cycle, highlighting the system's **adaptive capacity** rather than its limitations.
- **Deforestation and Soil Degradation**: Shift carbon cycles without breaking them, as the Earth rebalances through other reservoirs.

Action: Instead of attempting to control carbon directly, focus on maintaining the integrity of natural cycles, such as preserving forests, wetlands, and oceans, which act as the Earth's primary carbon participants.

2. Understanding the Role of Sinks

Forests, soil, and oceans are not separate entities but interconnected sinks that operate as part of a global system. When one reservoir is constrained, others adjust. Current atmospheric carbon levels reflect a redistribution within this balance.

- **Forest Loss**: While forest loss reduces terrestrial sequestration, the Earth compensates through other sinks like soil and oceans.
- **Soil and Oceans**: These reservoirs have vast capacities and dynamic feedback systems that can buffer carbon shifts.

TBT Insight: Carbon's movement between reservoirs mimics the principle of reciprocity in nature, where systems adapt to restore balance.

Action: Focus on protecting existing forests, preserving healthy soils, and safeguarding natural ecosystems like oceans and wetlands. Rather than imposing artificial solutions, allow these systems to function as intended, enhancing their natural capacity to maintain balance and resilience.

3. Human Activity as Part of the System

Revised Understanding of Carbon Fluctuations

Human activity represents a fraction of the total carbon output on Earth, contributing approximately 5% of the natural flux. The Earth's carbon system is vast and complex, with major contributions coming from natural processes like ocean-atmosphere exchange, vegetation cycles, and soil respiration. The discussion requires a shift in focus to reflect this broader understanding.

1. **Atmospheric Carbon Increase Reflects Systemic Redistribution:**
   - The 50% rise in atmospheric CO_2 (from 280 ppm to 420 ppm) aligns with changes in carbon sinks and reservoirs, not an outright "failure" of the system.
   - These fluctuations are adaptive responses, signaling shifts rather than failures.

2. **Natural Processes Dominate:**
   - Oceans, forests, and soil cycles process over 200 billion metric tons of carbon annually, while human emissions add only ~11.34 billion tons per year—a small perturbation in the grand scheme.
   - A slight imbalance in these larger systems can result in noticeable changes in atmospheric CO_2, independent of direct human emissions.

Removing Overemphasis on Human Impact

The narrative that ocean warming or deforestation drives the carbon imbalance oversimplifies a much more intricate system. These points require clarification:

- **Oceans:** While warming could hypothetically reduce carbon absorption, the temperature changes are minimal (~0.1°C per decade) and unlikely to significantly alter the massive flux between oceans and the atmosphere. Carbon flux is better explained as part of a natural, dynamic reallocation process.

- **Forests:** Deforestation impacts are localized. Globally, forests continue to sequester billions of tons of carbon annually, far outweighing the losses from deforestation. The capacity of the biosphere to absorb carbon remains robust.

Refocusing the Key Takeaway

Carbon fluctuations in the atmosphere are a symptom of natural reallocation, not human domination. Carbon acts as a signal—an indicator of the balance and flow within a vast, interconnected system.

The observed rise in atmospheric CO_2 is not a sign of systemic collapse caused by human activity but a reflection of natural processes at work. Human contributions, while real, are a fraction of the total flux. Carbon's movement highlights the Earth's adaptability and the interconnectedness of its systems. The solution lies in understanding these dynamics rather than imposing controls. Controls too have a consequence. (Below is data to back up my point.)

1. **Human Contribution (~5% of Total Carbon Flux):**

    - Human emissions of **~11.34 billion metric tons of carbon per year** are indeed small compared to the **~210 billion metric tons** cycled naturally through Earth's systems annually.

    - This fact is widely supported by data from organizations like NOAA and the IPCC, which outline natural and anthropogenic carbon fluxes.

2. **Atmospheric CO_2 Increase (50% Rise Since Pre-Industrial Levels):**

    - The increase from ~280 ppm (pre-industrial) to ~420 ppm today reflects a net accumulation of carbon in the atmosphere over centuries.

    - Natural sinks absorb around **55% of annual human emissions**, leaving **45% to remain in the atmosphere**, consistent with current scientific understanding.

3. **Ocean Absorption and Temperature Sensitivity:**

    - Oceans absorb **25% of annual human CO_2 emissions**, a fact well-documented in climate studies.

- While warming oceans could theoretically reduce this capacity slightly, the observed changes (~0.1°C per decade) have not resulted in significant disruptions to carbon sink efficiency. Most carbon absorption remains intact.

4. **Forests and Deforestation:**
    - Deforestation contributes **~1.5 billion metric tons of carbon annually**, but global forests still sequester **~7.6 billion metric tons of CO_2 annually**, maintaining their role as a net carbon sink.
    - These numbers align with global forest and land-use studies, which show that while deforestation impacts are significant locally, they do not negate forests' global sequestration capacity.

5. **Systemic Redistribution of Carbon:**
    - The concept of carbon being dynamically redistributed among reservoirs (atmosphere, oceans, soil, and biosphere) aligns with the fundamental principles of the carbon cycle.
    - The "excess" atmospheric CO_2 reflects ongoing adjustments within this system rather than outright failure.

Verified Conclusion:

This analysis demonstrates that human emissions are a **perturbation** within a much larger, self-regulating system. Atmospheric carbon changes are part of a **complex interplay of natural processes** and human activity. This perspective challenges the narrative of imminent catastrophe from human emissions alone and underscores the resilience of Earth's carbon cycle under current conditions.

4. Addressing Misalignment

The rise in atmospheric CO_2 reflects systemic redistribution within the carbon cycle, not a direct result of human activity alone. While human contributions, such as fossil fuel use and deforestation, play a role in carbon movement, they are minor compared to the overall scale of natural cycles.

- **Example:** The burning of fossil fuels releases carbon stored underground, contributing to a redistribution of carbon among reservoirs. However, this reallocation is part of the system's dynamic flow and does not constitute a fundamental imbalance.

TBT Insight: The perceived "excess carbon" is better understood as a signal of redistribution rather than an existential threat. Realigning human activity with natural cycles allows the system to self-regulate and maintain balance.

Action:

- Prioritize solutions that respect natural processes.
- Focus on preserving and enhancing natural systems, such as forests and soils, without attempting to control or overmanage them.
- Recognize that Earth's capacity for adaptation is vast, to which we only play a part.

5. A Call for Greater Understanding

The narrative of mass extinction due to carbon emissions is not supported by the numbers or the principles of TBT. The current shifts in carbon distribution reflect a natural process of adaptation. This is not a crisis but an opportunity to understand and participate in the Earth's rhythms rather than attempt to control them.

- **TBT Insight:** Earth's systems are self-regulating. Carbon serves as a teacher, reminding humanity to live within the natural laws that govern all systems.

Action:

1. **Stop viewing carbon as a problem to be solved and instead see it as a signal of the Earth's natural processes at work.**
2. **Promote education that highlights the scale, adaptability, and interconnectedness of Earth's systems, shifting the narrative from fear to understanding.**
3. **Recognize humanity's role as an integrated part of nature, capable of aligning with its cycles to foster balance and resilience.**

Key Takeaway

The carbon problem is not a crisis but an invitation to realign human systems with the universal principles of TBT. Carbon is not an adversary; it is a guide, signaling reciprocity, balance, and transformation. By embracing

TBT, humanity can transition from control to connection—letting the natural blueprint guide us toward a balanced and resilient future.

Conclusion: A New Perspective on Carbon

There is no real carbon crisis—only a profound need for deeper understanding. Too often, in our hubris, we assume all things are under our control. Through the lens of TBT, the solution lies not in controlling nature but in reconnecting with it. By aligning human activity with Earth's self-regulating systems, we can rediscover our place within the universal blueprint and allow the planet's inherent resilience to guide us toward balance and harmony.

I recognize that this perspective may challenge deeply held beliefs and may even feel offensive to some. Just as we explored faith in earlier chapters, we now confront another area of faith—science. Many people place their trust entirely in scientific consensus, which, though invaluable, is often subject to human interpretation and bias. My aim is not to overthrow anyone's truth but to illuminate a path toward truth itself.

I was made for truth. In this revelation of the *Transcendent Blueprint Theory*, I was given a framework to explore, understand, and address the challenges humanity faces. While I may not be the one to solve all problems, I firmly believe that this theory provides the keys to uncovering solutions by pointing us back to the universal principles that govern life.

With that, let us move from the atmospheric to the cellular, from the global to the personal. Let's see if TBT can guide us toward insights into one of humanity's most persistent challenges: cancer.

Chapter 13: Cancer Through the Lens of the Transcendent Blueprint Theory (TBT)

Disclaimer:

This book is intended for informational and contemplative purposes only. It is not a substitute for professional medical advice, diagnosis, or treatment. The concepts explored are theoretical frameworks based on the *Transcendent Blueprint Theory* (TBT) and are not intended as medical guidance. Always seek the advice of qualified healthcare providers with any questions about medical conditions or treatments. The author assumes no responsibility for the application or misinterpretation of the content herein.

Cancer, when viewed through the lens of TBT, emerges as more than a disease—it is a signal of systemic misalignment, reflecting universal principles of cycles, transformation, reciprocity, motion, and connection. By transcending conventional frameworks, TBT offers a different perspective on cancer, inviting us to explore its causes and solutions through the universal systems that sustain life.

1. Cancer as Disrupted Cycles

Cancer disrupts the natural cycles of growth, division, and death (apoptosis), leading to uncontrolled proliferation and imbalance.

- **TBT Insight**: Cancer represents a breakdown in the cyclical principle, where cells lose the ability to die with purpose, disrupting the natural rhythm of renewal and regeneration.
- **Actionable Focus:**
  - Investigate systemic factors that disrupt cellular cycles, such as chronic inflammation, oxidative stress, and hormonal imbalances.
  - Develop therapies targeting circadian rhythms or metabolic cycles to restore natural cellular behavior.

2. Cancer as a Breakdown in Reciprocity

In healthy systems, cells contribute to the body in exchange for resources. Cancer cells, however, hoard energy, grow at the expense of others, and disrupt the balance.

- **TBT Insight:** Cancer mirrors ecological imbalances, where overconsumption by one entity destabilizes the whole system. It is a response to disruption, not a rogue anomaly.
- **Actionable Focus:**
  - Reprogram cancer cells to restore cooperative behavior.
  - Explore therapies that block resource access to cancer cells without damaging surrounding healthy tissues.

3. Cancer and Transformation: The Double-Edged Sword

Cancer thrives on transformation, adapting to environmental stressors and mutating for survival, often at the expense of the host.

- **TBT Insight:** Transformation is a universal principle, and cancer exploits this adaptability. The issue lies in its misalignment with the host system's health.
- **Actionable Focus:**
  - Harness cancer's adaptability to induce less harmful behaviors or guide cells back into normal cycles.
  - Study how environmental and systemic factors trigger these transformations.

4. Motion and Friction in Cancer

Cancer is dynamic. Tumor cells move, invade new tissues, and evolve to resist treatments. Resistance—or friction—often strengthens cancer by fostering adaptation.

- **TBT Insight:** Cancer follows the principle of motion and friction. It thrives on resistance, using it to adapt and gain momentum, much like friction generates heat and energy.
- **Actionable Focus:**
  - Develop dynamic therapies that evolve alongside cancer, preventing it from gaining momentum.
  - Use cancer's own motion to deliver targeted treatments, turning its adaptability into a vulnerability.

5. Connection: Cancer's Interdependence

Cancer cannot exist in isolation. Tumors depend on their environment, manipulating blood vessels, immune cells, and surrounding tissues to sustain growth.

- **TBT Insight:** Cancer reflects the universal principle of connection. It thrives within the larger system, and its solution lies in understanding and disrupting these dependencies.
- **Actionable Focus:**
    - Target the tumor microenvironment holistically, rather than focusing solely on the cancer cells.
    - Enhance the immune system's natural ability to recognize and eliminate cancer.

6. Cancer as Evidence of Adaptation

Cancer adapts to survive, reflecting the TBT principle that systems under stress transform for survival, even in destructive ways.

- **TBT Insight:** Cancer's adaptability is not inherently destructive—it is a misaligned survival mechanism. The goal is to realign this adaptability with systemic health.
- **Actionable Focus:**
    - Investigate how stressors like environmental toxins or chronic inflammation trigger cancer's adaptive behaviors.
    - Develop therapies that guide cancer cells toward less harmful adaptations.

7. Healing Through Systemic Alignment

Cancer is not just a localized problem—it is a systemic issue influenced by the body's environment, energy flow, and external factors.

- *TBT Insight:* The solution lies in restoring systemic alignment. Addressing root causes of imbalance—poor nutrition, stress, and toxic environments—strengthens the body's natural ability to self-regulate and heal.
- **Actionable Focus:**

- Combine targeted treatments with holistic strategies like personalized nutrition, stress reduction, and detoxification.
- Design therapies that integrate cycles of renewal, reciprocity in energy flow, and connection.

Key Insight: Cancer as a Teacher, Not Just a Disease

Through TBT, cancer becomes a teacher—a reflection of disrupted principles and an opportunity to explore deeper truths about balance, adaptability, and interconnection. By learning from its behavior, we can uncover paths to healing that align with the universal blueprint.

Thought:

Cancer reveals both the fragility and resilience of life's systems. Through the principles of TBT, we can shift from thinking that revolves around fighting symptoms to understanding causes, unlocking solutions that honor the complexity of existence. The journey into cancer is not just the hope of eradicating disease but rediscovering the harmony that sustains life.

This lays the foundation for the next section on **Faith and Cancer**, exploring how intangible forces and unseen principles contribute to healing.

Faith and the Role of Intangible Forces in Cancer

1. **Faith as a Protective Influence:**
   - **TBT Insight:** Faith, much like motion or energy in physics, may operate as a stabilizing force. It aligns individuals with rhythms of healing and transformation, potentially reducing stress, fostering resilience, and enabling the body to self-regulate.
   - **Biological Lens:** Chronic stress and a lack of emotional or spiritual anchoring are linked to higher levels of inflammation, which is a known precursor to cancer. Could faith act as a buffer, mitigating these stress responses and maintaining systemic balance?
   - **Example:** Numerous studies suggest that individuals with strong faith or a sense of purpose experience better outcomes during illness, potentially due to lower cortisol levels and better immune function.
2. **Absence of Faith as a Disruptor:**

- **TBT Insight:** The absence of faith—or more broadly, disconnection from systems of trust, purpose, or belief—may act as a destabilizing factor. This could lead to misaligned energy flows, both emotionally and biologically, that weaken the system's defenses.

- **Biological Lens:** Depression, hopelessness, and prolonged stress correlate with weakened immune responses, increased oxidative damage, and impaired DNA repair mechanisms—all of which contribute to cancer development.

- **Question for Research:** Could the absence of faith or purpose correlate with biological markers of imbalance, making the body more vulnerable to cancerous transformations?

3. **The Unknown Influences: Physics and Beyond**

- **Laws of Physics:** Physics governs energy, motion, and entropy, all of which influence the body. For *example:*

- **Entropy:** Cancer may represent localized entropy, where cellular order breaks down. Faith or intentionality could hypothetically reduce this entropy, realigning the system.

- **Motion and Resistance:** Cells exist in constant motion, exchanging energy. Disruptions in these dynamics—perhaps due to unseen forces—could lead to the chaotic growth patterns seen in cancer.

- **Energy Flow in TBT:** The body is an interconnected system, much like the Earth's carbon cycle. Disruptions in energy flows (e.g., through poor nutrition, stress, or toxins) could mirror cancer's emergence as a physical manifestation of these misalignments.

4. **Other Intangible Influencers:**

- **Emotional and Social Connection:** Loneliness and isolation are associated with poorer health outcomes. Could a lack of connection, akin to the absence of faith, create conditions where cancer thrives?

- **Environmental Energies:** Some traditions, like Feng Shui or energy healing, posit that physical and emotional health are tied to energy fields around the body. Could disruptions in these unseen energies predispose individuals to cancer?

Key Insight: Cancer as a Mirror of Misalignment
Faith—or its absence—may play a role in cancer as part of a larger tapestry of influences, including unseen physical and emotional forces. Cancer could be viewed not as an isolated disease but as a signal of deeper systemic misalignment, both within the body and with the universal principles of TBT.

Next Steps in Exploration:
1. **Scientific Research:** Investigate correlations between faith, emotional resilience, and cancer outcomes, focusing on markers like inflammation, immune function, and oxidative stress.
2. **Physics-Based Analysis:** Examine how principles like entropy, energy flow, and motion might influence cancer development and progression.
3. **Holistic Healing:** Explore therapies that integrate faith, emotional support, and systemic alignment, blending biological, emotional, and spiritual interventions.

Closing Thought:
If faith holds the power to heal, it may reflect a profound alignment with the universal principles of TBT—cycles, motion, reciprocity, and transformation. Conversely, the absence of faith might not directly "cause" cancer but could create an environment of misalignment—a sink searching for a source or a host condition fostering imbalance. By recognizing the interplay of tangible and intangible forces, we unlock a deeper understanding and embrace a more holistic approach to preventing and addressing cancer.

Note: The concepts in this chapter are theoretical explorations based on the Transcendent Blueprint Theory (TBT). They are not intended as medical advice. Always consult with qualified healthcare providers for medical concerns.

Chapter 14: The Womb as a Blueprint for Life

The womb is not just a physical space; it is a **microcosm of universal principles**, a system where the laws of the universe reveal themselves in their purest form. In this sacred environment, life begins its journey, shaped by cycles, exchanges, and energies that mirror the design of the cosmos itself. It is here, in the cradle of potential, that the *Transcendent Blueprint Theory* **(TBT)** comes alive, offering profound insights into the nature of existence.

From the biological to the energetic, the womb encapsulates truths that scale infinitely outward. By examining the development of life in the womb, we can uncover lessons that transcend individual experience, connecting the smallest microcosm to the grandest universal design.

The Womb as a Microcosm

From the moment of conception, the womb becomes a universe in miniature. Within this self-contained system, every element—biological, environmental, emotional, and energetic—plays a role in shaping the developing child. These elements are not isolated; they interact, balance, and respond to one another, illustrating the TBT truth that all systems are shaped by cycles and flows.

  1. **Cycles of Growth and Rest:**
     - In the womb, life unfolds in rhythmic cycles—of rest and activity, growth and pause. These patterns reflect the universal law of rhythm found in TBT, where all systems move in recurring phases that sustain balance and progress.
     - Just as the Earth spins and seasons change, the womb operates within cycles that prepare the child for the rhythms of life.
  2. **Neutrality and Growth:**
     - The womb provides a neutral space, free from external disruption, allowing the child to grow according to its inherent blueprint. This reflects the TBT truth that neutrality fosters freedom and potential, creating conditions for unhindered growth.

Biological Foundations of Growth

The womb as a biological system demonstrates the TBT truth of **reciprocity**: nothing grows in isolation, and every action creates a reaction. The mother and child are in constant exchange, a perfect illustration of the dynamic balance of giving and receiving.

1. **Nutrition as Energy Flow:**
   - The mother provides nutrients that fuel the child's development, mirroring the cycles of energy flow seen in ecosystems. Energy moves through the mother's body to the child, following the path of least resistance to sustain life.

2. **Hormonal Signals as Guidance:**
   - Hormones like cortisol and oxytocin guide the child's growth, shaping its physical, emotional, and psychological framework. These signals adapt to environmental and internal conditions, reflecting the TBT truth of **contextual variability**: while the blueprint (DNA) is fixed, its expression adapts to the surrounding environment.

3. **The Genetic Blueprint:**
   - DNA provides the map for development, but external factors influence how this map is read. This emphasizes the TBT principle that fixed patterns (like DNA) interact dynamically with variable conditions to shape outcomes.

Environmental and External Influences

The womb does not exist in isolation but is shaped by the larger systems surrounding it. This reflects the TBT truth that **scale dependency** governs all systems: what happens on a small scale is influenced by larger forces, and vice versa.

1. **Stress and Resilience:**
   - External factors like pollution, noise, or maternal stress introduce friction into the system. While some stressors may harm development, others create resilience, illustrating the TBT truth of **motion and resistance**: friction shapes growth by strengthening the system's capacity to adapt.

2. **Motion and Rhythm:**
   - The mother's movements create rhythmic waves of energy that influence the child's development. These cycles of motion and

stillness prepare the child for the natural rhythms of life, aligning with the TBT truth of energy flow and rhythm.

Emotional and Psychological Energies

The womb is also an emotional and energetic environment, shaped by the mother's state of mind. This reveals the TBT truth of **vibrational influence**, where even intangible emotions create measurable impacts.

1. **Maternal Emotional State:**
   - Joy, fear, calmness, or anxiety are more than feelings—they are vibrations that influence the child's development. These energetic imprints show how subtle forces can shape outcomes, reflecting the TBT truth of **energy exchange**.

2. **Bonding and Connection:**
   - The child begins to sense the mother's presence through vibrations like her heartbeat and voice. This mirrors the TBT truth of **connection as foundational**, demonstrating that the bonds formed in the womb resonate throughout the child's existence.

Energetic and Spiritual Dimensions

Beyond the physical lies the realm of energy and spirit. The womb holds not just the child but the energy of its potential and purpose, revealing the TBT truth of **latent potential**.

1. **Energetic Imprints:**
   - Thoughts, intentions, and environmental vibrations leave energetic imprints on the child, shaping its early development. This reflects the TBT truth that energy moves across boundaries, influencing outcomes in unseen but lasting ways.

2. **Consciousness and Awareness:**
   - The question of when consciousness begins touches on the TBT truth of **universal intelligence**. Even before birth, the child is connected to the larger web of life, receiving and interpreting its first signals.

Universal Lessons from the Womb

The womb, though small in scale, holds truths that extend to the universe:

1. **Cycles Are Foundational:**

- From the phases of pregnancy to the rhythms of development, the womb operates within cycles that mirror the universal rhythms of nature and the cosmos.

2. **Reciprocity Sustains Growth:**
    - The child's development depends on the mother's energy, just as ecosystems thrive on mutual exchanges.

3. **Energy Flow Shapes Outcomes:**
    - Every aspect of the womb, from nutrients to emotions, follows the principles of energy flow, showing how systems are guided by the path of least resistance.

Conclusion: The Womb as a Universe

The womb is more than the beginning of life—it is a space where the universal truths of TBT manifest in their most intimate and profound forms. From cycles of growth to the flow of energy and the vibrations of connection, it reflects the patterns that shape all existence.

Through the lens of TBT, the womb is a reminder that what is true for the smallest microcosm is also true for the cosmos. It teaches us that life, at its core, is a balance of potential and expression, neutrality and motion, connection and independence.

In the end, the womb is not just a place where life begins—it is a blueprint for existence, a space where we remember the universal design that guides all things.

Part 4
The Future

Chapter 15: The Unique Promise of TBT

The **Transcendent Blueprint Theory (TBT)** is not merely a framework for understanding; it is a revolutionary lens through which humanity can reimagine existence. By observing and aligning with universal principles such as cycles, motion, reciprocity, and transformation, TBT provides the tools to revolutionize every aspect of life—medicine, technology, culture, and spirituality—while maintaining harmony with the natural systems that sustain the universe.

What Makes TBT Unique

1. Universal Laws as a Foundation

TBT identifies universal principles—cycles, reciprocity, motion, and transformation—not as abstract concepts but as observable laws found in every system. These principles bridge the tangible and intangible, uniting scientific and philosophical understanding.

- **Scientific Connection**: For example, in physics, the law of conservation of energy (energy cannot be created or destroyed, only transformed) mirrors TBT's emphasis on energy flow and reciprocity. This same principle governs ecosystems, where energy cycles between producers, consumers, and decomposers.
- **TBT Truth**: Every action and reaction contributes to balance, renewal, and growth across systems.

2. Transcendence Across Disciplines

TBT unifies diverse fields—biology, physics, philosophy, and faith—by demonstrating how universal principles manifest across systems. By connecting the tangible and intangible, TBT fosters innovation that transcends isolated disciplines.

- **Example in Biology**: The process of cellular respiration demonstrates cycles of energy transformation, paralleling the flow of nutrients in ecosystems and the cyclical nature of societal systems.
- **Example in Philosophy**: TBT echoes Aristotle's notion of the "unmoved mover," proposing that universal principles set systems into motion while remaining inherent and constant.

3. Practical Problem-Solving

TBT transforms interconnectedness from a popular concept into a tool for addressing critical challenges like climate change, disease, and perceived imbalance. By aligning human efforts with the natural blueprint, we avoid solutions that disrupt systems and instead create those that harmonize with them.

- **Carbon Cycles**: TBT reframes the carbon debate by emphasizing balance. For instance, planting trees to absorb carbon aligns with natural cycles, whereas artificial interventions disrupt them.
- **Medical Advancements**: Rather than treating symptoms in isolation, TBT encourages therapies that address systemic causes, such as promoting health to enhance immunity—an approach supported by emerging research in microbiome science.

4. Alignment Over Control

Conventional approaches often seek to control systems for specific outcomes, leading to unintended consequences. TBT advocates for alignment with universal principles, enabling harmony and resilience without overreach.

- **Historical Example:** The Green Revolution, while solving food scarcity, disrupted ecosystems and created dependencies on artificial fertilizers. A TBT approach would emphasize sustainable practices that implement freedoms for principles to operate.

A Vision for the Future

As a preeminent theory, TBT has the potential to redefine human understanding and progress across all disciplines. Imagine a future where:

- **Medicine**: Treatments align with the natural cycles of the human body, leveraging the body's regenerative capacities and rhythms.
- **Technology**: Innovations harmonize with nature, amplifying its strengths without disrupting balance. For instance, energy distribution systems inspired by mycorrhizal networks could

dynamically adjust to energy demand, reflecting the natural flow and balance seen in ecosystems.
- **Culture**: Societies celebrate humanity's role within the universal blueprint, fostering creativity, ethical decision-making, and voluntary alignment with life's natural rhythms.

In 100 years, I believe, TBT's principles will form the foundation for integrating science, philosophy, and spirituality into a unified framework for sustainable progress. This is not about quick fixes but a long-term vision for thriving through alignment.

The Risks of Misalignment
Revolutionary advancements, while transformative, often carry unintended consequences. TBT highlights the dangers of misalignment, as seen in historical examples:

- **Agricultural Revolution**: While increasing food production, it created population surges and ecological strain.
- **Industrial Age**: Rapid progress led to environmental degradation and in turn balance requirements by natural systems.

TBT Cautions:
- **In Medicine**: Breakthroughs should address root causes rather than creating dependencies (e.g., overuse of antibiotics disrupting microbiomes).
- **In Technology**: Innovations must enhance humanity's connection to natural cycles rather than introducing artificial systems that replace or degrade them.

Through systemic awareness, adaptability, and responsible innovation, humanity can ensure progress aligns with the rhythms of the universal blueprint.

TBT Truths for a New Age

1. Cycles Drive Progress
TBT reminds us that cycles—of birth, growth, decay, and renewal—are universal. Systems thrive when these cycles are respected:

- **Scientific Insight**: The hydrological cycle sustains ecosystems by distributing water. Similarly, societal cycles of innovation, stability, and renewal sustain human progress.

2. Reciprocity Sustains Life

Nothing in the universe exists in isolation. Systems thrive on reciprocity, where energy exchange fosters balance and growth:

- **Biological** *Example:* Symbiotic relationships, such as between pollinators and plants, exemplify reciprocity.
- **Cultural** *Example:* Societies flourish when collaboration and mutual benefit flourish through freedoms.

3. Transformation Enables Growth

Transformation is a universal principle, where challenges and disruptions drive growth and innovation. However, transformation must align with natural systems rather than impose artificial control:

- **Historical Example:** The harnessing of electricity revolutionized society, but much of this transformation—through dams, centralized grids, and other infrastructure—disrupted natural systems. These examples highlight the risks of overriding balance and neutrality in pursuit of progress.
- **TBT Insight:** The challenge of climate change is not in the changing climate itself but in humanity's misunderstanding of its role within natural cycles, particularly the carbon cycle. Carbon, when viewed through TBT, represents not a threat but a neutral element—a constant poised for dynamic balance. Neutrality and freedom within the carbon cycle create opportunities for natural pendulum swings, where excesses are absorbed and balanced over time. The true transformation lies in humanity recognizing its role as a participant, not a controller, allowing natural systems to self-regulate while fostering alignment rather than interference.

The Call to Action

As stewards of this revolutionary perspective, the responsibility is ours:

- **To Embrace Curiosity**: Explore TBT's potential with open minds and innovative thinking.

- **To Solve with Alignment**: Create solutions that reflect freedom which allow balance, ensuring harmony with natural cycles.
- **To Build for Generations**: Develop technologies, cultures, and systems that strengthen rather than disrupt the blueprint of life.

TBT is not just a framework—it is a revelation. It offers humanity a way to understand and reshape existence, unlocking the blueprint's capacity for harmony and innovation.

Conclusion: The Promise of TBT

The *Transcendent Blueprint Theory* is more than an intellectual framework; it is a transformative vision for the future. By aligning human systems with universal principles, TBT enables a harmonious balance between progress and sustainability. Its truths are not confined to one discipline or time—they are the eternal rhythms that guide existence.

Through TBT, we are not just solving problems; we are rediscovering the profound interconnectedness of life and the universe. This is the promise of TBT: a blueprint not just for survival, but for flourishing in harmony with the rhythms of creation itself.

Chapter 16: The Final Reflection

This journey through the **Transcendent Blueprint Theory (TBT)** has not merely explored knowledge—it has uncovered a unifying framework that bridges the deepest divides of science, spirituality, and human experience. It is a revelation of the profound design that governs existence, from the smallest particle to the vast expanse of the cosmos.

TBT is not just another theory. It is the first framework to align and reconcile **scientific principles, ancient wisdom, and spiritual understanding**—a groundbreaking step toward seeing the universe as it truly is: **interconnected, rhythmic, and alive with infinite potential**.

The Blueprint Unveiled

At its core, TBT illuminates patterns of life that have always existed but remained obscured, fragmented by disciplines that viewed the world in isolation. Through its principles, we see how the physical, biological, and intangible systems of existence—thought, energy, and matter—operate under the same **universal laws**.

1. **The Seed Law: Encoded potential at origin governs growth, transformation, and renewal.**
   - *Biblical Alignment:* "I have planted, Apollos watered; but God gave the increase" (1 Corinthians 3:6).
   - *Scientific Parallels:* DNA, galaxies, and ecosystems all begin with inherent potential that unfolds through cycles.

2. **Cycles and Renewal: Life thrives through continuous cycles— creation, breakdown, and rebirth.**
   - *Biblical Alignment:* "To everything there is a season, and a time to every purpose under heaven" (Ecclesiastes 3:1).
   - *Natural Law:* Forest fires clear the old to make way for new life; stars die to create elements that seed new systems.

3. **Reciprocity: The flow of energy sustains balance and connection.**
   - *Biblical Alignment:* "Give, and it shall be given unto you" (Luke 6:38).
   - *Scientific Parallels:* The conservation of energy ensures nothing is wasted, only transformed.

4. **The Law of Magnetism: Systems attract or repel based on alignment and motion.**
    - *Spiritual Echoes:* "As a man thinketh in his heart, so is he" (Proverbs 23:7).
    - *Scientific Principles:* Magnetic forces and gravity govern cosmic relationships, mirroring human dynamics of attraction and alignment.
5. **The Law of Neutral Freedom: True balance requires neutrality, a space where freedom and intention intersect.**
    - *Spiritual Alignment:* "Be still, and know that I am God" (Psalm 46:10).
    - *Philosophical Connection:* Balance does not suppress motion—it enables it, maintaining harmony in systems.
6. **The Law of Work: All systems require effort and action to manifest outcomes.**
    - *Biblical Alignment:* "Faith without works is dead" (James 2:26).
    - *Universal Parallels:* Motion, friction, and energy are required for creation, growth, and transformation.

A Unification of Truths

The power of the *Transcendent Blueprint Theory* lies in its ability to bring together **science**, **faith**, and **spirituality** in a way never seen before.

- **Science** reveals the principles—entropy, motion, magnetism, and cycles—that govern the natural and cosmic worlds.
- **The Bible** and other spiritual traditions echo these universal laws, articulating them through language of fire, seed, renewal, and balance.
- **Ancient Wisdom** intuited the same truths, viewing life as cyclical, interconnected, and purposeful.

For the first time, **TBT unifies these insights**, proving that they do not contradict one another but **complement and confirm** a single truth: **the universe operates through universal patterns that transcend boundaries, disciplines, and beliefs.**

The Breakdown Is Not the End

Heat, fire, and breakdown are central to this journey—forces that dissolve what is to prepare the way for what will be. The **fervent heat** described in **2 Peter 3:10** is not an end but a **refinement**:

- Stars die in fire to scatter the building blocks of life.
- Forests burn to clear space for new growth.
- Thought breaks down through friction to spark new ideas and progress.

In this pattern, we find a **greater cosmic cycle**: even the universe's heat death may not be final. Instead, it may mark a return to simplicity, a fertile ground for new beginnings.

Faith and science converge here, offering humanity the same truth—**destruction and renewal are two sides of the same universal blueprint**. The fire that breaks down also purifies, refines, and transforms.

The Call to Alignment

The *Transcendent Blueprint Theory* does more than reveal universal laws; it calls us to **live in alignment with them**. It challenges us to embrace the cycles, to honor reciprocity, and to recognize the potential within ourselves to **transform, refine, and create**.

We must ask:

- Are we resisting these patterns, or are we flowing with them?
- Are we choosing connection over division, harmony over chaos?
- Are we willing to align our actions, thoughts, and intentions with the rhythms that sustain all life?

When we live in alignment with the blueprint, we no longer see breakdown as an end but as an opportunity for transformation. We no longer fear cycles but embrace them as pathways to growth.

The Infinite Within Us

This is the ultimate revelation of the *Transcendent Blueprint Theory*:

- **We are not separate from the blueprint.**
- **We are the blueprint.**

Every action, every thought, and every intention flows within the same universal rhythm that pulses through the stars, the seeds, and the galaxies. We carry within us the same potential for renewal, balance, and transformation that governs all existence.

A Legacy of Harmony

This book has not been an end, but a beginning—a blueprint for understanding ourselves, the universe, and the infinite possibilities that await.

TBT stands as a unifying framework:

- It proves that science, spirituality, and ancient wisdom all point to the same universal laws.
- It shows us that destruction is not loss but transformation.
- It calls us to align our lives with the principles that govern the stars and the seeds alike.

Let this legacy be one of **harmony, courage, and renewal**—a movement where humanity thrives not by opposing life's forces but by aligning with them.

To understand the *Transcendent Blueprint Theory* **is to glimpse eternity. It is to see the infinite within the finite and the profound unity that connects us all.**

This is the truth, the invitation, and the promise of TBT:

We are not observers of the blueprint.

We are the blueprint.

Part 5
The End

Heat, Transformation, and the End of All Things

What book would be complete without a good close? And what theory complete without revealing an end. To a beginning. Science you can now see is evidence of our faith, and if so, is our God a consuming fire?

Ancient Truths Foreshadowing Universal Principles

"The heavens shall pass away with a great noise, and the elements shall melt with fervent heat..." — **2 Peter 3:10**

How did ancient wisdom, written in the language of prophecy and poetry, capture processes that modern science is only now beginning to understand? These words, penned thousands of years ago, carry striking parallels to the laws governing the cosmos—laws that reveal heat as both a destroyer and a catalyst for transformation.

At first glance, the imagery of melting elements and fervent heat may seem like an apocalyptic warning, a vision of finality. Yet beneath its surface lies a deeper, universal truth: **destruction is not the end but a necessary stage of renewal**. Heat, whether it burns through forests, consumes metals in fire, or dissolves stars into scattered elements, is part of a cycle that breaks down the old to prepare for what comes next.

This cycle—of breaking down, refining, and rebuilding—is not unique to the heavens or the earth but is written into the blueprint of all existence. It mirrors the natural order seen in forests after fire, in the purification of precious metals, and in the cosmic processes that drive the birth and death of stars.

Heat, then, is more than a force of destruction. It is the ultimate transformative agent, dissolving systems into their purest essence and scattering the seeds of new creation. The ancients may not have had our tools for measuring thermodynamics, entropy, or the fate of stars, but their wisdom grasped a universal principle: what burns away makes way for what is to come.

In this truth, Peter's words take on new significance—not as the final destruction of all things but as a revelation of the **transcendent cycle** embedded in the universe. A consuming fire does not erase; it transforms. It refines what remains and prepares the foundation for renewal, echoing the same patterns we see in nature, in cosmic law, and in the deepest structures of existence.

How did they know? Perhaps because truth—like heat—transcends time, boundaries, and disciplines, leaving its mark wherever we care to look.

Section 1: Heat as the Cosmic Decomposer

Decomposition is nature's way of returning systems to their simplest states—a breaking down that feeds what comes next. In the organic world, this role is fulfilled by bacteria and fungi:

- **Bacteria** break down flesh, dissolving complex tissues into their elemental components.
- **Fungi** decompose organic matter, transforming leaves, wood, and life's remnants into nutrients that fuel new growth.

These processes are not annihilation; they are transformation. They reduce systems to their essence, recycling energy and material into the next stage of life. On the grandest scale, **heat** plays a similar role in the cosmos. It is the **universal decomposer**, breaking down the fabric of stars, galaxies, and matter itself into their most basic forms.

Heat and Entropy: The Dissolution of Complexity

In the language of physics, entropy is the measure of disorder within a system. Over time, heat disperses energy, driving systems toward simplicity. Structures dissolve. Motion slows. Energy, which once fueled stars and life, spreads thin, becoming unusable.

This process is relentless:

- Stars burn through their fuel, collapsing into black holes or scattering their elements into space.
- Galaxies lose cohesion, their energy radiating into the void.
- On the smallest scales, heat erodes matter's structure, releasing it into forms too simple to sustain complexity.

This is the nature of **entropy**—a dispersal of energy that reduces systems to a state of balance.

Thermal Equilibrium: The Heat Death of the Universe

If entropy continues its work, the cosmos will reach a state scientists call **thermal equilibrium**. At this point, all energy will be evenly distributed, and all motion will cease:

- Stars will have burned out.
- Matter will have broken down into its simplest particles.
- Heat, once the driving force of transformation, will no longer flow, because there will be no difference in temperature between one point and another.

This vision—sometimes referred to as the **heat death of the universe**—is not fiery but cold, silent, and still. The universe will have broken down into a final state of perfect balance, stripped of complexity and motion.

Transformation, Not Destruction

At first glance, this process may appear like destruction—an undoing of everything the universe has built. But decomposition, whether organic or cosmic, is not the end. It is a return to the **fundamental building blocks** that fuel the next cycle.

In the *Transcendent Blueprint Theory*, heat's role aligns with the universal principles of **cycles** and **reciprocity**:

- Heat transforms systems, breaking down complexity into simplicity.

- Energy, though dispersed, is not lost. It is redistributed, preparing the foundation for new possibilities.

- In this sense, the breakdown becomes a prerequisite for renewal—a resetting of the system to its purest state.

Just as forests require fire to release seeds and regenerate, and fungi return life to the soil to nourish new growth, heat works on the cosmic scale to strip systems down, ensuring nothing is wasted.

Heat doesn't just destroy; it transforms. It is nature's ultimate decomposer, aligning perfectly with the cycles of breakdown and renewal written into the blueprint of all existence.

In both the smallest organisms and the largest galaxies, the story remains the same: **what breaks down prepares the way for what comes next**.

Section 2: The Fire of Stars and the Melting of Elements

"The elements shall melt with fervent heat..." — **2 Peter 3:10**

Peter's vision of melting elements aligns with a profound cosmic reality: the life and death of stars. At the heart of the universe's most powerful processes lies **fire**—not the earthly fire we know, but an unimaginable furnace of heat and energy that transforms matter itself.

Nuclear Fusion: The Fire that Builds and Breaks Down

In the core of every star, **nuclear fusion** takes place. Extreme heat and pressure force hydrogen atoms to fuse together, forming helium and releasing enormous amounts of energy in the process. This fire is the driving force behind the light and life we see across the cosmos.

Yet fusion does more than create light—it **transforms** elements:

- As stars evolve, fusion forges heavier elements, from carbon to iron.
- These elements, born in the intense heat of stellar furnaces, become the building blocks of planets, organic life, and entire systems.

This melting and transformation is written into the universe: heat breaks down simple atoms to create complex forms, embodying the **universal cycle of transformation**.

The Death of Stars: Fire's Final Act

Stars do not burn forever. When their fuel runs out, they die, and in their death throes, they release everything they created:

1. **Supernovae**: Massive stars explode, scattering their elements into space—carbon, oxygen, iron, and more. These elements are the **seeds of life**, forming planets, oceans, and even us.
2. **White Dwarfs and Neutron Stars**: Stars collapse into dense remnants, their heat and energy exhausted.
3. **Black Holes**: The most mysterious of all, black holes form when massive stars collapse under their own gravity. They become "cosmic furnaces," consuming matter and energy, transforming it into a state that defies our understanding.

A black hole, in essence, is the ultimate breakdown—a singularity where matter is compressed beyond recognition. Yet even here, the process may not be final. Some theories suggest black holes release energy back into the universe over vast timescales, continuing the cycle of transformation.

The Melting of Elements: A Universal Pattern

Peter's description of the "melting of elements" takes on a literal meaning in the fire of stars. Elements melt, fuse, and transform under the heat of nuclear fire. In their death, stars dismantle these elements and scatter them across the cosmos, where they become the building blocks of new systems.

This pattern mirrors the processes found everywhere:

- In organic systems, heat and fire decompose material, returning it to the soil to feed new growth.
- In cosmic systems, heat breaks down stars, releasing elements to seed new galaxies and life.

TBT Insight: Motion, Friction, and Transformation

The life and death of stars align perfectly with the principles of the **Transcendent Blueprint Theory:**

1. **Motion**: Stars are dynamic systems, constantly moving and transforming through fusion and radiation.
2. **Friction**: The intense pressure and heat within stars create the friction necessary for fusion—the process that builds and breaks elements.
3. **Transformation:** The fire of stars transforms matter, breaking it down and reshaping it into forms that feed the next cycle of creation.

The death of a star is not an end but a beginning. Its fire, though extinguished, leaves behind everything needed for new worlds to form. Heat, motion, and breakdown become the agents of renewal, ensuring that the universe—like all systems—follows the cycle of transformation encoded into its very fabric.

The fire of stars reveals a universal truth: breakdown is required for renewal. Melting, transforming, and releasing energy are not acts of destruction—they are the necessary steps that allow life, matter, and systems to rise again.

Peter's words, once seen as a final judgment, now resonate as a cosmic reality: *heat dissolves, but it also transforms.* The heavens, like the stars themselves, are not destroyed by fire—they are remade through it.

Section 3: Universal Patterns: The TBT Perspective

At every scale of existence—from the tiniest organism to the vast expanse of the cosmos—**heat** reveals itself as a fundamental force of transformation. In the framework of the *Transcendent Blueprint Theory* **(TBT)**, heat aligns perfectly with the universal principles of **cycles**, **reciprocity**, and the **Seed Law**, showing that destruction is not the end but an essential step in renewal.

Cycles: Heat as a Catalyst for Transformation

The universe operates in cycles—creation, dissolution, and rebirth—and *heat plays a key role in driving these processes.* These cycles are not only the driving force behind cosmic phenomena but are also fundamental to the continuous flow of energy and matter throughout existence.

- **Stars** burn through fuel, break down, and scatter their elements, seeding the next generation of stars and planets. This process is an ongoing cycle that allows new life to emerge from the death of the old. Without the breakdown of stars, new worlds and even the possibility for life would not exist.

- **Galaxies**, in their magnificent spirals, radiate energy. They evolve toward balance, shedding older, unused material while simultaneously giving way to new cosmic formations. This cycle of expansion and contraction, creation and destruction, mirrors the rhythms of life on Earth.

- On **Earth**, **fire** is the catalyst that transforms forests, breaking down old vegetation while preparing the soil for new growth. Forest fires clear the land, returning nutrients to the earth that allow new plant life to thrive. The same principle applies to ecosystems across the planet—destruction creates the necessary conditions for regeneration.

Heat, far from being a force of annihilation, serves as the catalyst for renewal within these cycles. It dissolves what is, allowing the emergence of what will be. It is not destruction for its own sake, but part of a **natural process** that ensures **transformation**.

Reciprocity: Energy Transforms, Nothing Is Wasted
In the universal balance, energy is never truly lost; it changes form. Heat, as the agent of breakdown, follows the principle of **reciprocity**. The breakdown of one system is balanced by the emergence of another, maintaining harmony in the universe.

- The **energy** released by dying stars becomes the building blocks of new celestial bodies. As stars collapse and explode, they scatter their elements across the universe, seeding the next generation of stars, planets, and even life itself. The death of stars facilitates the birth of new stars and the potential for new worlds.
- **Fire** that consumes a forest also nourishes the soil, clearing space for seeds to grow. The very destruction caused by fire creates fertile ground for new life to sprout. As the ashes enrich the soil, life's cycle continues, ensuring the flow of energy through the environment.
- Even in **decomposition**, energy is returned to the system, fostering the next generation of growth. Decay is not a final end but a transformation, a necessary step in the continuous process of renewal. Bacteria, fungi, and other microorganisms break down organic matter, releasing nutrients into the soil and maintaining balance within the ecosystem.

Reciprocity ensures that destruction feeds creation. The breakdown of one system is balanced by the rise of another, maintaining the universal equilibrium encoded in the blueprint of existence. Nothing is wasted; everything transforms.

Seed Law: Destruction Encodes the Potential for Rebirth
The **Seed Law** states that within every system, the potential for renewal is encoded at its origin. Even in cosmic destruction, this principle holds true. In every collapse, there is the potential for rebirth, making destruction and creation two sides of the same coin.

- When **stars** die, their remnants scatter into space, forming the raw material for new stars, planets, and even life itself. This perpetual cycle of death and rebirth extends to the very fabric of the universe, as new stars and planets emerge from the ashes of the old.

- **Forest fires** release seeds from trees that depend on the heat to germinate. The fire clears the ground, exposing the soil to the sunlight and preparing it for new life to take root. This process mirrors the regenerative forces at play in the cosmos, where destruction clears the way for new growth.

- **Decomposition** breaks down complex organic matter into simpler elements and nutrients, feeding the next generation of growth. Just as fire clears the land for new seeds, decomposition creates fertile soil for new plants and trees to grow, perpetuating the cycle of life.

Heat, then, is the agent that unlocks potential. It reduces systems to their essence, breaking down what was and preparing the ground for what comes next—whether physical, biological, or cosmic.

Natural Comparisons: Heat and Renewal

The principles seen in the cosmos mirror those found in the natural world. These cycles of breakdown and renewal are not limited to the stars or galaxies; they occur on Earth and within living organisms as well.

- **Forest Fires**: Fires consume forests, breaking down vegetation, and releasing essential nutrients back into the soil. The resulting ash enriches the soil, making it fertile for new life. The destruction of the forest is essential to its regeneration.

- **Decomposition**: Bacteria and fungi break down organic matter, returning nutrients to the soil. Through decomposition, life sustains itself. The decaying leaves of autumn nourish the soil, providing the foundation for spring's new growth. This continuous process ensures that energy and resources are never wasted.

- **The Cosmos**: Heat from dying stars scatters elements into space, providing the foundation for new celestial systems. The heat of a supernova transforms elements into the building blocks of new stars, planets, and possibly life. From the destruction of one star, a new world is born.

From **forests** to **galaxies**, heat strips away the old, reducing complexity and preparing the system for renewal. Destruction and creation exist in harmony, each dependent on the other to maintain balance.

TBT Insight: A Universal Pattern
In the *Transcendent Blueprint Theory* (TBT), heat becomes the agent of transformation, driving cycles of breakdown and renewal across all systems:

- **Cycles**: Destruction paves the way for transformation and rebirth. Heat creates the conditions for new beginnings by breaking down the old.
- **Reciprocity**: Energy doesn't vanish; it changes form, maintaining balance. What is lost in one place is renewed in another, ensuring that no energy is wasted.
- **Seed Law**: The potential for renewal is always present, encoded within destruction itself. Even in collapse, there is the promise of new life and new growth.

The universe, the Earth, and even life itself follow this pattern. What breaks down prepares the way for what comes next. Heat is not an enemy but a force that ensures systems stay in motion, that life continues, and that matter remains in a constant state of transformation.

Section 4: The Spiritual and the Scientific Unite
Across time and space, two voices—one from ancient texts and the other from modern science—speak a unified truth: the universe moves through cycles of breakdown and renewal. Whether described through the prophetic words of scripture or the language of thermodynamics, the pattern remains clear. The cosmos, like all systems, must pass through transformation, breaking down what exists to make way for what is yet to come.

The **Seed Law**, embedded within the *Transcendent Blueprint Theory*, suggests that the potential for renewal is encoded even in destruction. Just as the remnants of stars hold the building blocks of future systems, the final breakdown of the universe may prepare the ground for a new creation.

Faith as a Framework for Transformation
Faith offers humanity a way to make sense of this breakdown—a lens through which destruction is not feared but understood as part of a

THE TRANSCENDENT BLUEPRINT

transformative process. Peter's words remind us that fire does not erase; it refines. It melts away what is temporary, revealing what is eternal.

In the same way, science shows us that nothing in the universe is wasted:

- Energy does not disappear; it transforms.
- Matter breaks down, but it feeds the cycles of creation.

Science and faith converge on a profound truth—the cosmos' breakdown is not meaningless. It is a necessary step toward something deeper, something more refined, and perhaps something *eternal*.

In the end, both perspectives affirm a cycle that transcends destruction. The fire that melts also prepares. The breakdown that dissolves also renews. Whether seen through the eyes of faith or science, the message is the same:

What ends is not truly gone—it is transformed, refined, and prepared for what comes next.

Closing Reflection: The Fire That Renews

The universe speaks through patterns—cycles of creation, destruction, and renewal. Heat is both the destroyer and the catalyst for transformation, its presence marking the final stage of all things and the beginning of something new.

Peter's words, *"The elements shall melt with fervent heat"* (2 Peter 3:10), echo the cosmic truth we see unfolding in stars, galaxies, and the laws of thermodynamics. Matter breaks down, systems dissolve, and energy scatters—but this is not the end. It is the transformation encoded within the **Seed Law**, where destruction paves the way for renewal.

Even in faith, the imagery of heat holds profound significance. The Bible describes **God as a consuming fire** (Deuteronomy 4:24; Hebrews 12:29)—a fire that purifies, consumes imperfection, and transforms the unworthy into something refined. Just as gold is purified by fire, and organic matter is broken down to feed new life, so too does this fire work on cosmic and spiritual levels.

The universe does not waste. Nothing burns without purpose. Heat breaks things down not into oblivion but into simplicity, preparing the foundation for what comes next. The stars that dissolve in fire scatter the very elements that birth new worlds. The forests consumed by flames release seeds and nutrients to rebuild ecosystems. Destruction, in its truest form, is creation delayed—a transformation held in tension.

When viewed through the lens of the *Transcendent Blueprint Theory*, the consuming fire becomes a universal force:

- A purifier of systems.
- A catalyst for cycles.
- A manifestation of reciprocity, where what breaks down feeds what rises anew.

If the universe itself is consumed by heat, as Peter foresaw, it may not signal a tragic end but the unfolding of another cycle—cosmic renewal, driven by the same principles that sustain life on every scale. Heat strips away the old to make way for the pure, the true, and the essential.

Whether seen through the cosmos or the divine, fire reminds us that **nothing is wasted**—what breaks down will rise again, reborn in purpose, because the blueprint of existence is infinite.

Biblical References to Heat and Fire

The Bible consistently uses **heat and fire** as symbols of transformation, judgment, and renewal:

1. God as a Consuming Fire:

- *"For the Lord your God is a consuming fire, a jealous God."* — **Deuteronomy 4:24**
- *"For our God is a consuming fire."* — **Hebrews 12:29**
- This fire consumes imperfection, refines purity, and aligns with the cosmic role of heat as a force that breaks down systems to prepare for something better.

2. The Refining Fire:

- *"He will sit as a refiner and purifier of silver; he will purify the Levites and refine them like gold and silver."* — **Malachi 3:3**
- Heat purifies precious metals, separating impurities to leave only what is valuable. This mirrors how cosmic heat refines the universe, breaking it down into its fundamental building blocks.

3. **The Fire of Judgment and Renewal**:
   - *"The fire will test the quality of each person's work."* — **1 Corinthians 3:13**
   - Fire is used to test, reveal, and cleanse, aligning with heat's role in nature and the cosmos to expose what is essential and prepare for new creation.

4. **The Fiery Furnace of Transformation**:
   - The story of Shadrach, Meshach, and Abednego (Daniel 3:17-27) highlights fire as a trial that transforms faith but does not consume the faithful. Fire becomes a metaphor for enduring and emerging stronger.

5. **Tongues of Fire and the Spirit**:
   - *"They saw what seemed to be tongues of fire that separated and came to rest on each of them."* — **Acts 2:3**
   - Fire here represents a transformative, purifying energy—echoing heat's role in the physical world as a catalyst for change.

The Unifying Insight

From the cosmos to scripture, heat and fire reveal themselves as universal forces of transformation. They consume, refine, and break down, yet always in service of something greater: purity, growth, or rebirth. This pattern, encoded into the universe, aligns seamlessly with the Transcendent Blueprint Theory and the Seed Law.

Peter saw fire as the end—but perhaps, as the blueprint shows us, it is only the beginning. A consuming fire breaks down what is, but only to reveal what *will be*.

And thus ends this book.